My Story

SFC Patrick W. Marshall, USA Ret.
Pioneer, California

My Story
Copyright © 2013 by Patrick W. Marshall

ISBN: 978-1618290632

10 9 8 7 6 5 4 3 2

My Story

DEDICATED TO

MY WIFE---MY LOVE---MY PARTNER

****Sandy****

At 2,000 feet, close the doors and relax or write letters.

Introduction

This is my story, not written as a novel with a story line, but rather a collection of thoughts and remembrances. Mostly of my time in the U. S. Army from 1962- 1999, 27 years total, with an 11 year break after 1970. And then back in for 18 years in the Army Reserve. These are my views and not necessarily those of the U.S. Army.

My children and grandchildren have asked me what the medals on my uniform are for and why I got them and especially the Distinguished Flying Cross. I guess it would be best to start from the beginning.

Never in my wildest dreams did I ever plan to be a soldier or make a career out of the U.S. Army, but I did. Growing up I had my dog Sport, my fishing pole, my 22 rifle and pistol, a good horse and the whole Sierra National Forest as my back yard. What more does a boy really need? Nothing! One of my friends and I would saddle up just before Christmas and ride up one of the logging roads in the snow looking for Christmas trees. In the summer my sister Connie and I would sometimes pack a lunch and follow the logging railroad right of ways into the high country or the 'Central California Stock Driveway', riding until early afternoon and then coming home before mom got worried about us. Sometimes picking apples in one of the old orchards planted by the early settlers and long abandoned.

In the beginning, my beginning anyway, I was born in my grandmother's house in the small coastal town of Cambria, CA just south of Hearst Castle. 18 Years later after high school, I was happily working part time with my father in Oakhurst, CA, he was a contractor, and I was working as high country guide and packer in the summer and shoeing horses in between time. 'Smoke' Floyd Adams had taught me to shoe horses. He was a WWII airborne veteran and an American Indian. We spent a lot of time together training horses in the Oakhurst, Bass Lake area as well as shoeing. If you have seen the movie; "Horse Whisperer," well Smoke could go up to a strange horse and breathe into its nostrils and talk softly and the horse would follow him anywhere. One day we went to an Indian Rancheria in South Fork, just east of North Fork and South of Bass Lake, to look at a wild stallion. This horse was not people friendly and it was kept in a small corral not more than 20' X 20'. This is I n the dead of winter and after negotiating on a price of some cheap whiskey, a six pack of beer and a few dollars; Smoke climbed into that corral with ankle deep mud, walked right up to that horse and breathed on him and said a few kind words. His breath had to be really bad by this time with the booze and all. He then put his belt around the horse neck and led him up and into the back of his pickup; it had stock sides on it. This horse had never been in a trailer or a truck before. After paying the Indians for the horse, I had to drive home as Smoke, having drunk a bit with the Indians, was not in any shape to do much of anything but sleep, and with his cattle dog "Sis" curled up beside him on the truck seat we left. The Indians just stood there, not believing what they saw him do.

A few years later in the Army, my dad sent me a letter that 'Smoke' had died in his sleep. Must have drunk too much and fell asleep with a 'smoke' in his hand and burned his trailer up. Drunk or sober he would take off his shirt and give it to you if you needed it.

The man didn't have a mean bone in his body. And, on top of that he was a wonderful animal artist, sometimes paying his bar bill with sketches on napkins, place mats or any thing available at the time. The letters that I got from him were always a combination of words and pictures. In one saloon outside of Fresno, in the foothills, hung one of his paintings called "Range Romance" he had painted it in the late 40's or early 50's. It was of a Proud Jackass, a mustang mare and her mule foul in the desert, enough about 'OL' Smoke, could write a story just about him. If I new where his daughter was I would try and do that just for her.

It is ironic, and it wasn't until recently that I found out the meaning of my family name, 'Marshall'. It means 'The Keeper of The Horse," a position that gained respect in medieval times as a trainer, handler and later as a person that, also, shoes horses.

How I wished that I could have raised my children in the atmosphere that I grew up in, working outdoors as I did without the video games and other nonsense that is so prevalent these days. Every one tells me that the kids need those games and cell phones to become computer literate. Hog wash! Children need to go out and get dirty, making their own fun as they go. And while I am on that subject;

boys were boys and their parents would never let them go to school in raged clothing and pants down around their butts. Girls were proud to be girls and dressed the part in dresses and skirts. No shorts, no tattoos, and the only body piercings were in their ears. Kids need to be kids first. Computers will come when needed.

During deer season boys often carried rifles in their cars or trucks to school so they could go hunting after school. Even our Vise Principle, Mr. Franz carried his rifle and hunted around our high school in the afternoon. There was no big panic about weapons near the schools in that day. Friend Jim and I would sometimes ride in his car to school and bag a duck or goose on the way. Not wanting the bird to spoil we would drive out in the country and cook it over an open fire. The next day, we forged a note to Mr. Franz from our parents about our absence.

Chapter One: You're In the Army Now

Anyway, at age 21 I received a letter from The President of the United States, and it said "Greetings" you are to report for a pre-induction physical at the Fresno induction center. Several years later I would end up back at that location in a different capacity. If any of you have ever taken that physical, you know that if you can stand and walk 'you're in the army now and not behind the plow' so, not wanting to be drafted, I took a bunch of tests and enlisted for Army Aviation on the 29th day January in 1962. After basic training at Fort Ord, Company C, 11th Battle Group, 3rd Brigade. I went to Fort Rucker, Alabama for basic aircraft mechanics school and then on to Headquarters 18th Airborne Corps aviation section at Fort Bragg North Carolina. At Bragg I was in charge of Aviation Supply and I got involved in developing methods of rappelling from a helicopter and climbing out of tall pines and back into a helicopter after a simulated rescue or action against the rebels in South Vietnam. The commander of the 18th Airborne Corps and Fort Bragg, Lt Gen Howze, not sure of the spelling any more, would come to the airfield with his ideas. I would come up with a design, get his approval and then build it. The paper trail needed to requisition anything in the Army was anywhere from one to two weeks long. The General had his aid prepare a letter for me so that I could walk a requisition through in one day. Short cuts were not the army way but when he wanted something done, it was now and not next week. This did not make me popular at any of the supply locations or with the folks that processed the final paperwork.

Lt. Chapman, myself and our rappelling gear with a UH-1B and the folding box 'platform.'

We had a team of four enlisted men and Lt. Chapman as our team leader. We put on several demonstrations for Congressmen and others. The Lt. was the first to rappel out of the Huey and I was the last. The Lt. led the other three on their assigned mission while I, using a preselected tall tree, climbed the tree, cut out the top and prepared it for us to use as a platform to get back into the helicopter after the mission was completed.

Because the rappelling ropes would get tangled in the trees if just dropped out the door of the Huey, I designed a reel devise that the first person out the door carried behind him to deploy the ropes. Then each man held the ropes for the next as a safety measure and to keep them from getting tangling in the trees. We were training an engineer unit to rappel into the forest to cut a clearing big enough for a Huey to land and to build a log platform for it to land on. One of the young engineers panicked and grabbed on to a tree part of the way down, refusing to let go. His rope was getting tangled in the tree at the same time. I had to rappel down along side him and convince him to let go and then go down with him slowly, untangling his rope as he went. This incident led to the design of the reel device and a lot more training before anyone was permitted to rappel. Speaking of training, my airfield commander taught me in one easy lesson to rappel. He hooked up and out the door he went on 120-foot ropes and then stood on the ground yelling up at me to follow. Couldn't hear a word he said but I'm sure that they were not kind words as I hesitated to jump off the skids of the helicopter. Our team enjoyed rappelling so much that we used to rappel off the airfield control tower until we were told to stop because we were wearing out too many ropes.

The General not being a young man had me design and build a folding box, big enough for a man to stand in and hang it in the top of the tree. From there we could climb up a rigid ladder attached to the floor of the helicopter and hanging a couple feet below the skids on the left side. Gen Howze was scheduled to get the command in Viet Nam and he had lots of ideas to use over there. Didn't happen, Lt Gen. Westmoreland got the command instead.

Climbing a pre-selected tree at Fort Bragg.

Back into the chopper from the top of the tree.

A demonstration for Army and Congressional brass.

This was in a very wet, swampy area. Unfortunately the wind was too strong that day and we could not complete the demonstration. The photo was as close as the helicopter could get to the tree. Not good enough to complete the mission safely.

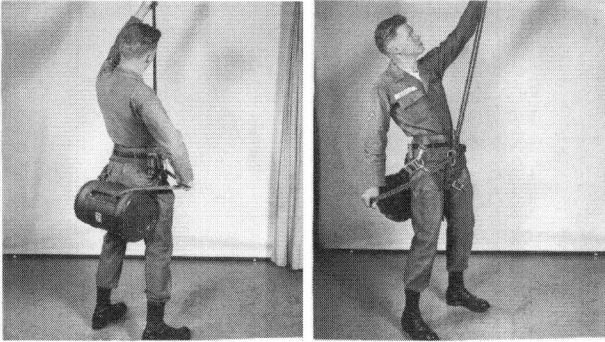

The reel device to keep the rope from tangling in the trees. Had to put brake shoes in it to keep the rope from backlashing like a fishing reel.

While we were doing all this an order came down one day to get our vehicles from the motor pool, load all of our spare parts and other go to war equipment, get them weighed and make a diagram of everything's location, then unload and put everything away. About a week later President Kennedy had his speech one evening about the missiles in Cuba. After the speech no one said a word, just got our weapons, field gear and extra cloths, drew our vehicles and went to the airfield and loaded up. Later that night we were assigned to cargo aircraft that would be taking us to the Marine Base in Cuba. No one was happy when the invasion was called off a few days latter. The 82 Airborne Division had gone to Florida to pack enough parachutes so they could all make the jump down the middle of the Island while Marines and Army hit every beach around the island. Rumor was that the 82nd after packing all their chutes stored them in a R.R. Box Car and that the boxcar disappeared with all the chutes in it. I believe that the 101st Air born was in on the invasion also. The 18th Airborne Corps was to run the show from

"Gitmo" as it is called now. Aside from what the media has reported and the official government reports say; we could have knocked out all the missiles in one night and taken the Island in no more than a week or so. Contrary to what you all were told, the missiles were not taken out and returned to Russia. The president made some kind of a deal with the Russians and then lied to the public. "The idea being; what we don't know won't hurt us." It could very well have hurt the entire country. You ask me how I know, well you don't have a need to know, just accept that it happened that way. More about this later.

Back to Lt. Gen Howze; he was a real gentleman and it was a pleasure to work with him. He was my first personal contact with a general officer. He had formed the Howze board that was ultimately responsible for the Army's acceptance and use of the UH-1 helicopters from Bell Helicopter Co. and the beginning of the Air Assault concept as it was to be later developed in Vietnam and at Fort Benning, GA by the 11th Air Assault Div.

Somewhere during all of this I was given a different job designator, Crash Rescue Specialist, (fireman on an airfield). This was done in hopes of getting me promoted to Specialist 4th class. Not being airborne and in the army's airborne headquarters and having the general ask my company commander why I was still a Private First Class; my goose was cooked. You see the company commander was only a captain and couldn't get promoted and my aviation section leader was a major. The captain was told by Gen. Howze to retire or get put back to the enlisted rank that he had in Korea before his field promotion. Why, because he made no effort to continue his education, he became non-promotable.

Remember when all the trouble started in Mississippi? The General told the Military Police Commander to send troops down there and get things settled down before they got out of hand. The Provost Marshal sent them down there without ammunition for their guns. Smart move! Wasn't long before the blacks and whites figured it out and started throwing stones and other things at the MP's. Howze, when he found out chewed out the Provost Marshal. (MP Commander) He then sent his driver to draw ammunition and we flew him down there personally so that he could issue ammo to the MP's. The General did this in the town square in front of the whole town. No more serious trouble with the natives. At least for a while!

While at Fort Bragg I went to visit the post stables on a day off. Boy what a cushy job the guys there had. Uniforms were worn only on payday and for special occasions. One of the Colonels on post had a couple horses for his wife and daughter at the stables. A GI that claimed to be a Ferrier had gotten a transfer to the stables and had almost crippled the hoses by his lack of ability at hoof trimming and shoeing. The Warrant Officer in charge tried to get me transferred to special services and the stable but my Commander refused to approve it; I was however able to help them get their horses hooves back into correct shape on my days off. The uniform of the day at the stables was clean jeans, a western shirt and boots. Had I received the transfer I might still be there. Also, I didn't know that a person could enlist to be a horse handler and go to school at Cal

Poly, CA and be taught to shoe horses. I guess it just was not meant to be.

Ft. Bragg had many lakes and ponds for fishing. You do not want to go fishing alone at night. The Water Moccasins will swim up and get into your boat. We always fished with a lantern hanging over the bow and one person in the boat with a machete and a flashlight. Right off the airfield was a nice picnic area by a small lake that families liked to use on weekends. It had to be put off limits midway through the summer because of the snakes moving in from the surrounding ponds and lakes. The forest is full of Cotton Mouths and if anyone ever tells you that rattle snakes do not swim, do not believe it, water may not be their favorite, but they will swim if it is the shortest distance to where they are going.

Chapter Two: Off to War

In June of 1963 I received orders to go to Viet Nam. Passports were still required to go there as all of the Americans in the country at that time were advisors and not supposed to be "Fighters." I even had to get a Vietnamese drivers license to drive American military vehicles when I got there. I went home on leave before going over there and my father having been in the Marines twice had some great advice for me. "Son," he said, "get to know the people, what their needs are and their hopes. Do not ever think that you are better than they are and put yourself above them. Don't go around flashing a wad of money and learn their customs and make friends." So after a week at the Oakland Army terminal and on the 4th of July 1963 I landed in Saigon South Viet Nam and spent a night there in a tent-house by a cemetery.

The soldier that showed me where I was to sleep told me to be sure and put all of my stuff up on another bunk. It rained that night and about 6 inches of rain came through the tent. In the morning I was flown up to a small city in the mountains, Pleiku. I was the senior firefighter on the airfield at Camp Holloway and did not have a clue how to operate the fire truck. Fortunately, there was one school-trained fire fighter there and he helped me with operating the pumps and other equipment on the fire truck. About once a week we would empty the truck's water tank into a large shallow concrete tank that the house girls did the laundry in. That was the extent of my operating the fire pump on the truck. The rest of the time was sitting and reading in the truck. Out of boredom, I did however get a piece of the local hardwood and carve a figure of a squat pregnant looking woman that some of the Vietnamese thought was some kind of a god. I had seen one similar to it in a book about Africa.

Camp Holloway main gate.

With a group of friends in Pleiku.

Some time later I was sitting in the fire truck by the control tower when another soldier walked by with his radio on, and said that President Kennedy had been killed. My first reaction was to get angry, thinking he was joking. Later that day, I went downtown and many of the older people were in shock and crying, thinking that we would abandon them after the loss of our President. They did not understand that our government is set up for just such emergencies.

About that time we had a visit from Special Forces soldiers that had just returned from the field. They had gone by an orphanage run by Vietnamese Catholic Nuns. The V.C. had just paid them a visit. After tying the nuns to the porch posts, and making them watch, they proceeded to rape all of the little girls that were not infants and crush their nipples with pliers, castrate the boys and other things that you may not want to hear about. As our soldiers arrived the nuns were just managing to get loose and many of the children were lying on the ground bleeding to death. I bet you never read about that over here! This was a real wakeup about the cruelty of the enemy.

One morning as I was sitting in my fire truck, a Captain that I knew from the airfield at Fort Bragg came by and seeing me asked what I was doing in the fire truck. He got me transferred to the aviation maintenance detachment on the airfield. A short time later I turned in my M1 carbine and we were issued new weapons, M-14's, with ammunition but no magazines? A week latter our Battalion commander decided that we needed to have a stateside type of a full field inspection with our weapons all stripped down, magazines empty, we had just received them. We were having this full field inspection while we were at work with only one soldier guarding our equipment in each hooch.

Brig. Gen Stillwell (Jumping Joe) came in for an unannounced visit. When he saw all of our equipment laid out on our bunks, he went to the Headquarters building and set off the alarm for being attacked. It was panic time as we all ran from our work locations for our hooch's and grabbed our gear and headed for our assigned foxholes. Only then did we have time to stop, assemble our rifles and load our magazines with very nervous fingers, thinking we were about to be attacked. Gen. Stillwell announced that it was not a real attack, called a battalion formation, chewed out the battalion commander in front of the troops for his stupidity and gave us the rest of the day off to get everything back in order.

In Pleiku they were still flying the old CH-21 (Flying Banana) from Korean War vintage. It was armed with two 30-caliber machine guns and the crew chiefs and gunners also had what they called 'Skippy Bombs'. Empty glass peanut butter jars with a fragmentation hand grenade, pin pulled inside and the lid screwed back on. If you saw bad guys, you simply threw them a 'Skippy Bomb' and when the glass broke, the handle came off and in a couple seconds "BOOM" This predated the armed gun ships that we got later. The CH-21's had wood blades and if it rained you had to find a place to land pretty quick as they would take on water and get out of balance. Bullet holes in the blades were patched with a wood plug and pieces of wood glued over the holes, sanded down and heavily waxed.

On a lighter side, while I was stationed in Pleiku I bought some beautiful white and some blue silk and sent it home for my mom and sister and they made Sister Connie a prom dress from the white silk that was that envy of the other girls at our high school. We never had much money and her showing up in that silk dress blew everyone away. Almost forgot, bought a beautiful pair of Whit Sapphire earrings for no one in particular and a star ruby ring for myself. Had the earrings redone for my wife at Christmas, 44 years later in 2007.

About that time, Bell Helicopter Company set up a school in Qui Nhon City for people switching to Huey's from the CH-21 and I was chosen to attend. The training I received there, prepared me to become a crew chief later. During my time at Qui Nhon the Vietnamese President was overthrown and we had no idea what would happen to us. We drew our weapons, ammo and a 50 Caliber machine gun that no one knew how to set up and placed it on the runway behind some sand bags. Fortunately nothing happened and the South Vietnamese were happy about the change of control and his demise.

While at school in Qui Nhon I met a Vietnamese couple that owned a beach restaurant and bar. I went to their home for dinner a couple times while I was there. They had a German shepherd that guarded the beach place during the day. If you said, "get the VC" the dog would crawl from table to table and tree to tree growling as he did so, and then come and roll over onto his back to get his tummy scratched. Qui Nhon had beaches that would rival anything in Southern California and about a mile out of town was a seafood restaurant owned by a French family, Oh! What food; had some excellent high-class restaurants in town also.

Restaurant on the beach and the 'VC' sniffing dog.

One other incident comes to mind while in Pleiku. There was a large, 3-hole, outhouse that set on a concrete slab. An elderly Vietnamese man had been hired to keep it clean and neat. He was very conscientious about his job. Everyone thought that he was just a poor elderly person taking care of a large family and everyone was always giving him gifts for the family…it turned out that he was a Col. in the V.C. local brigade and got the job to gather information against us. Up to that point we thought only the local barbers that worked on post were the V.C. intelligence agents. So go figure! On every assignment I had in Vietnam, if we were attacked at night, in the morning you would find at least one of the post barbers shot dead in the perimeter wire.

Our supply sergeant was selling all of the small sizes of uniforms and boots to a Vietnamese in town along with ammunition. Someone rigged a case of 45 Ammunition with a hand grenade that would explode when the can was opened. One local store, the owner and the sergeant went off together in a rather large explosion. Never got confirmation of this but we did get a new supply sergeant soon after.

Back then we generally had an unwritten agreement with the V.C. that the town was off limits to hostilities. After all it was a small city by our standards with only so many restaurants, bars and shops.

We had to share. What a concept. The guy next to you that bought you a beer or that you bought one for might have been one of the bad guys that you were chasing during the day. What a way to run a war, although I might say it was better than the way it became after we started bringing in multitudes of soldiers. A short time before I got there, on one convoy from Qui Nhon to Pleiku the V.C. told the American Commander not to bring any South Vietnamese security along. The Vietnamese insisted and the convoy was attacked in route. All of the Vietnamese soldiers were killed and one American Lt. that started to get out of his jeep had a couple rounds hit the road by his boots. Getting back into the jeep, all was fine because the Americans did not fire on the V.C. The V.C. then melted into the forest and the convoy continued to Pleiku without their South Vietnamese Army escort.

Occasionally one of the flight crews would get permission from the local Buddhist monks to shoot a young water buffalo and would usually bag a local deer at the same time. We would have a BBQ and anyone with talent would put on a show for us. All of the Vietnamese employees would be invited and for most of them a BBQ was a unique experience but generally enjoyed by all. They had no idea about our methods of cooking over on open pit with charcoal and BBQ sauce, not to mention the beans and other goodies that Americans are noted for. For most of them, cooking over an open fire was nothing special.

Deer for dinner.

Party time! You got talent, use it!

Young BBQ water buffalo and venison. Yum, Yum.

Not long after receiving our first Huey gunships a pilot was flying to close too the ground as he flew over a ridge just outside of town, crashed, rolling the chopper up in a ball with the tail boom tucked under the nose. With my experience rappelling in the states I had became one of the rescue team. The crash was in the open but we went in anyway to secure the area. No survivors.

Before

After

Each hooch had a "House Girl" that did your laundry, made the beds and polished you boots and shoes. The girl in my hooch was not much over four feet tall and her stepsister was the house girl in the next hooch. Her stepsister's husband was a Chinese pharmacist in Pleiku and they invited me to come to the pharmacy for tea one afternoon. If you have never been in a Chinese Pharmacy you have missed a treat. The entire back wall behind the counter was covered with small drawers and shelves containing herbal medicines and God only knows what else. Interesting smells as you walked the length of the front counter. An elderly woman came in and the Pharmacist took her into the back for an examination. Coming back out, he took scraps of newspaper and put on them, all sorts of things from the drawers and shelves, wrapping each into a small bundle and giving instruction on its use. Almost forgot the tea. An extra cup was poured but not served. This was significant but I cannot remember why. The tea was much like the 'Green Tea' that is so popular today. Very nice family, they invited me to have supper them, but it was getting late and their home was in the part of town that was not considered safe for Americans. Instead I went down the street and had some fried rice and Ba Mui Ba, the local Vietnamese beer. We learned to drink it warm or with ice. If the ice was frozen with a Coke bottle cap in it, it was made from potable water.

Otherwise you had your beer warm. The French had taught them how to make beer. Better than a Bud Light! Ba Mui Ba is number "33" in Vietnamese. Ba being three and Mui ten. (3X10+3=33)

We had one young man in the maintenance detachment that got a "Dear John" from his wife. The next day, after not showing up for work, he was found in his hooch passed out and almost died. He had bought a bottle of Scotch and drank all of it and passed out. The medics were barely able to get to Saigon and the hospital to save his life. Never had much use for a woman that would do that. Maybe that is why I did not get married until I went off active duty.

In Vietnam the original inhabitants that live in the mountains were called Mountain Yards, (Spelling). The Vietnamese looked them down on. One day on trash detail I took the trash truck to the dump to unload. The Yards, as we called them, were there looking for anything usable that we discarded and the Vietnamese would try and chase them away so that they could sell our cast-offs. The Vietnamese learned that when I was driving the truck, the "Yards" had first pick. We had taken them out of their villages and relocated them close to town for their 'safety'. They traditionally lived in rice straw homes, built on stilts and grew wild rice in clearings that they made in the mountains. Special Forces enlisted many of them as scouts and fighters. On night missions they would often put away their rifles and resort to using crossbows with hardened bamboo arrows. They were as stealthy as the traditional American Indian and could put an arrow in your eye at 50 paces with no trouble. Unfortunately the V.C. forced many of them into labor with the threat to rape and kill their women if they did not cooperate. The Yards were a simple people that just wanted to be left alone. The men, unless in town, wore only a loincloth and maybe a shirt. The women only wore a wrap around skirt of beautifully hand woven material. They would only wear a shirt if coming into town and the men would usually put on a pair of trousers or shorts as they approached town. Most of the Vietnamese would make fun of them for not being civilized (wearing clothing).

Chapter Three: UTT Helicopter Co. Saigon

A lot more happened at Pleiku but we need to move on. Because I wasn't authorized as a fireman in the maintenance detachment I came down on orders to be transferred to Saigon and the "Utility Tactical Transportation Helicopter Company" or UTT Helicopter Company as we were known. The commander of UTT asked me what I was trained for and I told him aviation maintenance. He sent me to their maintenance section to help train some of the new mechanics how to perform part of the 100-hour inspection procedures on their HU1-B helicopters.

UTT was the first and only helicopter company of gun ships in the Army and for that matter the world at that time. Not long after arriving in Saigon the weapons officer took me to their range to learn how to be a door gunner. That involved sitting in the door of a Huey and shooting at empty fuel barrels on the ground as you flew past them with an M-14 in the full automatic position. As the chopper broke left or right you were expected to have a full magazine to cover the belly of the ship when the mounted guns could not. About a week later I was called into the Commander's office and asked if I would like to be the crew chief on his Huey. His crew chief had been seriously wounded the day before and had to be airlifted to Japan and the Army hospital there. He had been hit in the shoulder and throat by 50 caliber bullets. Saw him a year later at Fort

Benning. He was trying to get back to the war after that surgery that fixed it so that he could talk again. That was the last time I saw him, don't know if he made it. Hope so! He really wanted to get back over there.

SABER '6'and co-pilot.

So starts my adventures as a crew chief. I drew my rifle from the arms locker, a 2-gallon jug of coffee and a 2-gallon jug of ice water and headed for the flight line. The helicopter had been shot up the day before and the maintenance supervisor showed me how to stop drill the cracks radiating from the bullet holes and put tape over them, Green tape! A lot like very strong duct tape but green and not shiny. He said if there is no structural damage we do not do sheet metal repairs until it's time for scheduled maintenance.

That was my first day as a crew chief on "Saber Six" the Company Commanders Huey and my first flight. The

CO was always the first to leave base and the last to return, quite often on our 20-minute fuel reserve. That way he knew what his people would be getting into and to be sure that they all got back safely if possible. We returned to the area that they had been to the previous day and encountered some of the same group from the day before. To get the picture right, an armed Huey had two mounted machine guns on each side mounted on pillions that extended out about three feet. These guns could be fired by the co-pilot who had a gun site attached above the windshield on his side of the helicopter. With the site he could move the guns and fire left or right or up and down. The pilot could only fire the guns from the stowed position, or straight forward. Also attached to each pillion was a stack of four double rocket launch tubes.

We had four machine guns and sixteen 1.75-inch rockets fixed to the helicopter, the crew chief in the left cargo door and the gunner in the right cargo door. The crew chief and gunner were armed with M-14 automatic rifles and twenty magazines each with twenty rounds each. We also carried a Vietnamese observer to 'identify' the enemy. His real job was to keep our magazines loaded from about 500 or so loose rounds that we carried for that purpose. Here we are my first day out and a V.C. took a shot at us from on my gunner's side. The V.C. missed and my gunner missed him. The CO put the Huey in a tight left turn and told me to get him. The VC after firing one more shot then tried to cross a small canal to escape. I fired a short three round burst from about 15 to 20 feet and watched his head explode. The rest of the day was relatively uneventful as I recall. Most of the V.C. had left the area or melted back into the populace.

Why you ask, have I described in detail the death of another human? You must remember that he was someone's son, a brother and possibly a husband and father. I believe that it is necessary that we remember our actions even in war. And that we never become comfortable or complacent in taking another human life. We were required to keep a body count on a daily basis. OK, turn in the numbers at the end of the mission and then forget it! In today's world mass shootings are almost commonplace in the civilian sector. I blame a lot of that on the violent video games, TV shows and movies that our young people watch. Our society has become so insensitive to violence that they have no concept of death and see that as an answer to most of their problems.

In combat if you take too much time to consider the consequences of your actions you may be coming home in a body bag. The enemy usually gets the first shot. It's you job not to give him a second chance. But never ever learn to enjoy it. If you learn to enjoy killing, then you have become a murderer and are no longer a soldier. Unfortunately, too many young men don't have the spiritual upbringing needed to understand that and they spend too much time playing video games that glorify killing.

My father used to send me newspaper articles about what was going on in Viet Nam and I would re-write the articles if the action happened near me and send them back to him. He would then post them on the local town bulletin board in Oakhurst for every one to read the truth. Never have understood why the media seems to be so biased. Joseph Galloway's report about his time with Col. Moore in the 1st Cav Div. is the only truthful thing I have ever read about the war over there. That comes with my second tour to Vietnam.

Sitting in the door of a Huey is usually safer than being a ground pounder. Fortunately the VC hadn't yet learned to lead a moving target. We had an advantage in that respect and we usually saw who we were shooting at rather than just movement in the brush and trees. Seeing your target clearly also means that you are aware of the number of people that you shot. We used only tracer ammunition and became very good at hitting whatever with the least amount of rounds expended. The tracers were also good for burning hooch's and anything else flammable.

The enemy often had tunnel entrances in their hooch's (homes) and would run to them after shooting at us. Being constructed mainly of rice straw they burned rapidly with just a few tracer rounds fired into them.

We flew missions 5-6 days a week in support of troops all over the country, usually South Vietnamese forces (ARVN). On one occasion we were supporting an attack on a suspected VC stronghold in a village South of Saigon. After the troop helicopters, CH-21's, dropped off the ARVN soldiers, they came under heavy fire and hunkered down behind some rice paddy dikes and started cooking their lunch rather than attack the village and be shot at. Typical of the ARVN soldiers at that time. Being along the coast, one of our Huey's went down, crashing into the sea. All aboard were lost including a British Air Force Wing Commander flying as an observer. I don't remember if they were shot down or just crashed. As we attacked the village one of our ships was making a gun run on a machine gun position in a camouflaged bunker. Off to the left front was an old abandoned French Catholic church and not knowing that they had a 50-caliber machine gun in the open door that ship was shot down by it and crashed into the machine gun bunker. Everyone died. My pilot came around and saturated the front of the church with machine gun fire while one of our platoon leaders with a 48-rocket configuration came around to the side and destroyed the rest of the church. Never could find many parts of that 50 or the people that were using it. That village turned out to be the best-defended position that we ever went up against. Loosing 2 Huey's, 8 Americans, 1 British officer and 2 ARVN's. In the 8 months I was there this was the worst loss that our company suffered in one day's action. After this action, a mockup of an aircraft cockpit was made and every one had to get out of the seatbelt and shoulder harness while under water in a pool at the officers quarters in Saigon fully clothed and with helmet and flack vest on. Turned out that a couple of the guys couldn't swim and had to be taught the basics first. We also started carrying life preservers if we were operating close to the sea.

Not much left after crashing into the gun position in the camouflaged bunker.

When the battle was over and we had taken the village, you wouldn't believe the prisoners that were taken. All of the men were out attacking an ARVN outpost in another district. The toughest battle we ever fought was against the women of the village.

Men will back off and fight another day if the odds are against them. The women will stand and defend their homes to the last. All of the prisoners were ether wounded or out of ammo and could not fight any longer. Don't ever let anyone tell you that women are the weaker sex. Mess with mama's house and you are in deep trouble.

Remember I mentioned Spiritual training. (Christian training) Another time we were out for an entire day and could not find any of the supposed enemy that was reported. The crew chief and gunner sat leaning out of the Huey rifle in hand and with a white smoke grenade ready to throw. White smoke is an indicator that you have drawn fire and the rest of your flight will take the appropriate action. Back to the story…we were flying and nothing was happening, or so I thought. I suddenly felt a sharp pain in my back causing me to sit upright instead of leaning out the door. As sudden as it came, it went away and I thought no more of it. After landing back in Saigon my gunner came around my side to dismount and put away the machine guns. His face turned white and he pointed to the ceiling of the Huey right above my head. We never heard any shots fired all day but there was a bullet hole above me that should have gone through my chin and out the top of my head. It could have only happened when that sharp pain made me sit up straight.

Don't ever laugh if someone tells you about Guardian Angels. They are there, they are real, and they will take care of you if God still has a mission for you here on this physical earth.

Another memorable mission…after getting a distress call from an American Advisor with the ARVN and hearing him get shot over the radio, we rushed to the area, about ten minutes from where we were. As we approached the location we could see soldiers in tan uniforms all over the area. This was the first time we came up against North Vietnam Regular Army (NVA). I saw the red star on their helmets as we approached and yelled at my pilot. He didn't pay attention to me and continued to make a low pass over them, waving as he did so. The next thing we knew, they opened up on us with a captured 50-caliber machine gun. Remember my Angel… two rounds came through the center of the Huey. One of them between my gunner and me putting aluminum fragments from the floor into our legs, going through the four trays of machine gun ammo under our seats and stopping only after hitting a hardened aluminum structural member behind us. It only stopped because it was a tracer round and not a solid one. The co-pilot tried to lay down suppressive fire but the guns jammed because of the damage to the trays of ammo. Flying up to 2,000 feet, my gunner and I cut the seat out of the way and bypassed the bad ammo; we climbed out on the pillions and recharged the guns. After test firing by the co-pilot the pilot refused to re-engage the enemy.

After reporting his failure to engage, our commander relieved him of flying a gun ship and put him in charge of POL (Fuel). I still have that expended round that I recovered from my ship. Had it been a solid round it would have taken out my engine and possibly the transmission or to put it simply, we would have been casualties, the kind in body bags.

One day we went out as usual looking for something or someone to shoot at and found lots of reasons for shooting. We had come into an ARVN post to re-arm and refuel. After leaving I did not re-zip my flack vest all the way up, was a really hot and humid day. Tucking my cap inside of my flack vest we went out looking for more trouble. All of the caps in the UTT were made at Cheap Charlie's Taylor shop in Saigon with our rank and unit embroidered on the front and our name on the back. As we were flying and doing our thing, my cap flew out and the last I saw of it, it was falling toward the village that we were attacking. Getting home that night I was still fuming about having to order a new hat as I walked into my hooch. Looking to my bed and on the pillow was my cap. It was their way of letting me know that they knew who I was and where I lived. You see all of the flight crews in the UTT had wanted posters on them around the country. As a crew chief I was worth between $500 and a $1,000.00 to them dead or alive. You must realize that the average farmer only made about $15.00 a year in cash. Most of goods and services they needed were bargained for with rice and vegetables. A thousand dollars would be a retirement that far exceeded their wildest dreams. Anyway the pucker power set in and I did not go into town for a couple days. Some VC must have brought it back by bus and then sent it on the base with one of many VC that worked there or may have intimidated some Vietnamese employee to deliver it.

Another early morning adventure, right at the crack of dawn we were looking for trouble and flying about ten feet above some scrub palms, when crossing a small clearing, and being half asleep, I was awakened by orange flames going past my face. Muzzle flashes. My wingman said I threw a smoke grenade so fast that I hit one of the VC with it. Our platoon leader was flying at about 500 feet that morning directing our action. He came down and opened up with his 48 rockets and the small clearing became a larger one. No prisoners were taken. I recall staying awake for the rest of the day without any problems. Once again, thank you to my angel.

Every day that we flew was an experience that I would not want to do over, yet I would not give it up for anything. We had a job to do and we did it very well. Supporting American Special Forces one weekend we stopped at the Catecka Tea and Rubber plantation owned by a French couple. He really liked Roi-Tan cigars, so every time we went up there one of the pilots would bring him a box. Their Daughter was home from a Swiss school and brought a friend with her. One of our pilots spoke enough French to communicate with the couple and we usually drank tea, and those that played Canasta would enjoy a game with them in the Gazebo by the pool. They had to pay the VC not to burn them out at night and the South Vietnamese forces to not throw them out in the daytime. The man had an old airplane maybe a Piper Cub that was their only real connection with outside world, so they had a dirt runway by the house. One of our Huey's went out on a mission. We usually fly in pairs, but this time only one went out. And when they came back the two young women were in their bikinis by the pool in front yard. The pilots got distracted as they hovered up the plantations dirt runway creating a dust cloud and landed with the left skid gear in a drainage ditch and rolled over. I'm not sure how they explained destroying a Huey, but somehow they did not have to pay for it. I would return to that tea plantation on my second tour with the 1st Cavalry Division.

Pay attention to flying and not the girls!

On a lighter note, the chapel on our base in Saigon had a real rat problem so on a trip south one day some of our officers made a trade with some special forces for a Python. After putting a six-foot snake under the building the rat problem was eaten. After that someone would periodically purchase a chicken or two to feed the snake. Speaking of rats, back in Pleiku there was a mongoose in our company area, a wild one, and it would move from hooch to hooch catching the rats. No! You never petted the mongoose. He would take your finger off if you tried. He just sort of tolerated us living in his source of food, the hooch's.

Over there you tucked in your mosquito nets under your mattress, not so much to keep the mosquitoes away but to keep from being bit by a rat. Upon rolling against the netting, you just automatically moved away. They would bite through the net if you were touching it. Still in Pleiku one of our pilots came back from a flight tired, walked into his hooch in time to see a monkey pick up a picture of his wife and slam it to the floor. One round from his 45 and the monkey was all over the wall and floor. Panic set in for a few minutes thinking that we were being attacked with the sound of gunfire. And while I am on the subject of monkeys, another monkey that was a pet was constantly getting into trouble, took a tumble out of a CH-21 at about 500 feet with a flare chute attached. As soon as the shock wore off he tried to climb the shrouds and streamed in. He was not missed a whole lot by us.

Back to Saigon, I was suffering from boils in a rather uncomfortable place on my body and lying on my bunk with my feet up on my footlocker. It was on the foot of my bed so I could elevate my feet and legs. The boils came a week or so after having gangrene in my left arm in two places from rocket fuel burns. I was busy engaging an enemy when the pilot announced that he was firing rockets. I couldn't close the door with someone trying to shoot me so the solid fuel propellant from the rockets exhaust burned me. Anyway, looking across the hooch there was a rat walking across a shelf with an apple in its mouth, pushing another apple to the end of the shelf. I threw a boot at it and it just looked at me and continued on in its endeavor. Had to use my entrenching tool to kill it, the rats over there are about the size of a small cat and most of them have fairly long hair. The antibiotic that the flight surgeon put me on had a negative reaction and my vision went blurry for a short time so I ended up in an Air Force dispensary for a few days on different antibiotics until the boils went away. They came back briefly on my next two tours over there. Seems that once you get a staph infection, it stays with you.

One of the men in our company brought home a small bear cub, the kind with a white V under its neck. Another man had a monkey that liked to bite. About once a week it was necessary to double up your fist and hit the monkey in the head as hard as you could to stop him from biting. The monkey kept trying to drown the bear in their water bowl. The bear grew up and held the monkey down until he drowned. Good riddance. The bear was a better pet anyway. Every one brought him plenty of fruits and vegetables from the mess hall.

On one mission we were supporting an ARVN assault on a village. The ARVN soldiers were being transported in CH-21 helicopters, Korean War vintage, and in that hot climate couldn't get more than about 50-75 feet off the ground with a full load. We were flying at maybe 150 feet placing suppressive fire on both sides of their flight making sure that no VC were hiding in the brush and trees. All of a sudden an old B-26 bomber from WWII vintage flown by American pilots flew under us to attack the village with napalm. They were lucky that we were not firing down as they flew under us but their luck soon ran out. Should not put a jet jockey in an old time bomber. Upon his arrival at the village the pilot pulled the nose straight up and released his bombs just like he would from a jet. As soon as he pulled into a steep climb, both wings came off. Don't know if they ever found the pilots. The old B-26 was not designed for that kind of stress on its wings. Even if it was, it was too old for that kind of maneuver. Between the bombs and the plane crashing, the village was pretty well destroyed.

Did I mention; Perry Masson came to visit for a few days? His people just about had a cow when he asked to go out on a mission with us. Can't remember if it happened.

Raymond Burr visiting the UTT Hel. Co.

Another person that I met over there was Brig Gen. Stillwell. Remember I mentioned him from my time in Pleiku, he used to come out to the airfield and fly as a gunner about once a month just to get out of the office and feel like a real soldier again. The general and I were sitting in the door of my helicopter one day waiting for the pilots and making jokes about how to end the war. My suggestion was to hire a bunch of crop dusters and have them spray Poison Oak oil concentrate over the enemy and then stand by with a field hospital with lots of calamine lotion. He thought that was a good idea.

One of his was to bring a bunch of rattlesnakes and cottonmouth water moccasins and turn them loose in the Delta. There is just about every other poisonous snake known in that country but none that liked water like the moccasins or that gave warning like a rattler. The last time I saw the Gen. we were on a plane coming back to the states and some of the guys in my unit came by the plane wearing VC outfits and carrying a San Pan that they had captured. Later they shipped it home to the Gen. Unfortunately a year or so later he was flying out in the Pacific and his plane was lost. Never did know how, or find any trace of him. Great lose to the Army and this country. He was known as 'Jumping Joe Stillwell' and his father was 'Vinegar Joe Stillwell' from WWII fame. He was a real soldier and not a politician, preferred fatigues over a class "A" uniform and a rifle over a desk.

On another mission we meet a 12-year-old boy that had just finished his first patrol with the civilian defense forces. Sad story…two years earlier the VC had come to his home and demanded his family's entire rice crop. His parents hid him and his older brother and begged to be left enough to eat and plant the next crop. They were killed and the crop taken. The next year with help from neighbors and family the two brothers got in another crop. The same thing happened and this time they killed his older brother. After that he joined the Civil Defense Forces (CDF) and on the day that we met him he went on his first patrol. He shot and killed a VC that jumped out on the trail in front of him. Remember this boy is only 12 years old and he is proud that he killed someone. Not a good way to start your life. We had one "slick ship", with a vulture painted on the side; no guns and it would bring us food and ammo in the field. It had just arrived so we gave the boy a sandwich and an apple. He had never seen an apple before or had a lunch-meat and cheese sandwich.

After his first kill, his first sandwich and an apple.

Some of us went South to another camp to help them transition from CH-21's to Huey's. This camp had such a rat problem that an ongoing contest awarded men for the most and the largest rats caught.

One of the crew chiefs there sent home for an air rifle so he could lie in his bunk shooting rats. The airfield was an old French facility and occasionally one of the land mines on one side of the runway, left from the French would just cook off, sending people into a panic. We went out one day with one of the crew chief trainees as my gunner. We got into a fight and the bolt on my rifle broke, I took my gunners rifle and the same thing happened to his. After returning to base and rearming we went out again, this time with borrowed rifles.

The one I borrowed broke also; the next time we came in I did not go back out. Its one thing to be able to return fire on the enemy and another to hear them firing at you and not be able do anything about it. Later that evening my chopper went out on a search and destroy mission with a crew from that base only. They flew a little too low and came back in with palm fronds stuck in the machinegun barrels and a broken chin bubble. The chin bubble is the Plexiglas bubble in front of, and below the pilot's feet. The co-pilots leg was all beat up from contact with part of a palm tree and Plexiglas shards. When we left that base we left our helicopters for them and came back to Saigon and new choppers.

It was somewhere about this time that I was notified that my father had suffered a severe heart attack and I came home on emergency leave. He was in the VA hospital in Fresno, CA. The hospital let me have a room for a couple dollars a night to pay for the linens and I stayed for about a week, until he was recovering almost enough to go home. He became so excited and talkative with me home on emergency leave that the hospital had to put him in a private room. They were afraid that he might have another heart attack if he would not stop talking to everyone that he saw about my being home from the war.

After a couple weeks home it was time to head back to Oakland Army terminal and five days waiting for a flight back to Saigon. I was traveling in Class B uniform so they could not put me on any details other than headcount at the mess hall.

On arriving back in Viet Nam, the new Huey that I got before the emergency was being taken care of by a new crew chief in the company. I told the maintenance officer not to take it away from him if he was doing a good job and that I could wait for the next one that came in. Needing some flight time to keep qualified for flight pay one of the other crew chiefs let me take one of his flights. When we flew we had a belt around our waist with about a 6-foot strap that snapped to one of the cargo rings in the floor, a helmet and gloves. Not having any of my gear with me he lent me his helmet and rifle, I went on the mission, no gloves, no belt and strap. It was a very busy day expending our load of munitions several times. We were making a rocket run on hooch flying at about 20 feet in elevation when our observer pulled the pin on a white smoke grenade and within a few seconds the interior of the chopper was a total white out. The pilots unable to see pulled up hoping that we were going to miss the hooch and not crash. I tried to lunge for the grenade but was restricted by the seat belt. By the time I got it loose and laid my rifle on the deck, the grenade was smoking vigorously and had reached temperature of a few hundred degrees. Grabbing it with my bare hands and juggling back and forth because of the heat I was able to throw it out the door. For this action I was put in for a Purple Heart but because I did not go to the flight surgeon with my burns, my commander instead put me in for the Distinguished Flying Cross. I didn't know it until I was in the 11th Air Assault Div at Ft Benning, GA. My Battalion Commander LTC Jack Cranford, 227th Assault Helicopter Battalion in 1965 presented the award to me.

General Orders number 115, 29 Jan 1965
Award of the Distinguished Flying Cross

The citation reads: On 16 June 1964, "For heroism while participating in aerial flight: Specialist Marshall distinguished himself by heroic action on 16 Jun 1964, in the Republic of Vietnam. On this date Specialist Marshall was serving as crew chief of an armed UH-1B helicopter in support of Vietnamese popular forces against the Viet Cong. While on a firing pass, a smoke grenade exploded within the helicopter, obscuring the vision of the crew. Specialist Marshall, with complete disregard for his own personal safety, unfastened his safety belt, dropped to the floor, grabbed the grenade with his bare hands, and then crawled to the open door and threw the grenade from the aircraft. Regaining his rifle, he then directed accurate rifle fire into the Viet Cong though hindered by burned hand and arms, and eyes filled with smoke. His prompt and courageous action saved the lives of the entire crew and reflects great credit upon himself and the military service. By direction of the President under the provisions of the Act of Congress, approved 2 July 1926, and department of the Army Message 941895, dated 22 October 1963."

Some times things are slow to catch up with you in the Army. I had received several Clusters to my Air Medal while in Saigon but didn't get the basic Air Medal until Maj. Gen Kinnard of 11th Air Assault Div pinned it on me at a parade in Benning almost a year later. At the time that I received the DFC I was told that if I stayed in the Army and retired, I would be eligible for a pay increase based on that award. I was told after retirement, by a board of review that the act wasn't heroic enough. I guess we could have all died instead.

Interestingly, the members of our unit didn't consider minor wounds as worthy of a Purple Heart and therefore, we often did not even go into see the flight surgeon and if you did, and it wasn't serious nothing came of the incident. In later years, over there, I have seen soldiers get the award for being scratched and cut on thorny bamboo. Such is life!

So there you have it! But this is not end of the story. I had extended in Viet Nam to get a promotion to Sp 5th Class before the emergency leave, so I still had a couple months to go. Like I said before, we flew 5-6 days a week unless we were down for scheduled maintenance or to repair battle damage.

We really got into trouble one day with most of the helicopters getting shot up. My CO's chopper had so many bullet holes no one could believe that he could keep it in the air. After returning to a friendly base his crew chief was unconsciously rearming it and started to rub his butt, and someone noticed blood on his trouser leg. He had a bullet fragment in his butt so the CO gave him a shot of Morphine and told him to lie on the grass and be still. He was up and trying to return to what he had been doing. He started limping and complaining about his foot. Removing his boot the CO found another round lodged in the bottom of his foot and gave him another shot and ordered him to be still; "I don't want to find anymore holes in your body" he said. The helicopter was so badly shot up that it could not be flown back to Saigon. My helicopter was one of the few that didn't get hit that day.

My angel is still at work. About this same time I remember watching a Huey lose its tail boom in flight at about 1,000 feet. As it spiraled down its load of Vietnamese troops were being thrown out both sides by centrifugal force. The crew may have survived but the pilot out of habit pulled up on the collective just before touch down. The co-pilot did get out but went back to help the crew chief and about that time the ship exploded. He inhaled fire and died a short time later. The pilot was also a fixed-wing pilot and may have tried a high-speed pedal turn out of habit. The mounting bolts for the tail boom had ripped through the heavy aluminum mounts on the fuselage. Nothing else seemed logical to cause the accident. The pedals in a helicopter control the tail rotor and any sudden application at high speed to one of the pedals causes undue stress on the aircraft frame.

Yes, I know, I have not said anything about Saigon. It was a beautiful city and I will never call it Ho Chi Minh City. There were lots of really good restaurants and theaters, if you didn't mind daubed in Vietnamese and French sub titles. I had a very nice young woman named Kim as a girl friend. Her husband had been a Captain in the ARVN forces and had been killed by his own men when he ordered them to attack a strong point in an ambush that had been set for them. Most of his company died also because they did not obey orders. My friend, Kim, had been to school in Dalate and could speak and write English very well. After her husbands death she received about ten dollars a month from the government and found a job as a manicurist in the barbershop I went to in Saigon. The barber thought I should get to know her and he trusted that I would be kind to her after her lose. Her mother never knew that she dated an American. She was very old fashioned and traditional. One evening, Kim and I went to see "Cat on a Hot Tin Roof" in Saigon with daubed in French and Vietnamese sub-titles. She kept trying to explain the story to me. That was the third time I had seen the movie. Usually we would go out to dinner in one of the fine restaurants that were in Saigon and once she took me to the Saigon Zoo. Kim had a five-year-old son that I never got to meet. She was afraid that he would tell his grandmother. Kim was a very special lady and I often wondered if she met the right person and remarried. I certainly hope so. She was afraid to come to this country and deal with the cultural differences. We corresponded for almost a year after I came back to the states. At that point in time neither Sandy nor I knew that we would meet, fall in love and get married when I was getting out of the active army at Ft. Ord a few years later.

Kim at the Saigon Zoo.

Chapter Four: Home and a New Assignment 11th ASLT Div

After that trip to Viet Nam and a month's leave at home where, after his heart attack my dad had become a locksmith and part time conductor on a tourist railroad that ran on part of the old Madera Sugar Pine Lumber Co. right of way, my dad asked me to paint a picture of the engine that used to run on that location with sugar pine logs to the mill. I started the painting but didn't get back to it until I came back from Vietnam the last time. My son Matthew now has the painting, still unfinished. After that I was stationed at Ft. Benning, GA in the 11th Air Assault Div. Test. The old 11th Airborne colors were reactivated into a Division to test the Air-Mobil concept. Gee, I thought that was what we had been doing in Vietnam. As an experienced crew chief in Co C 227th Aslt. Hel Company I was made Platoon Sgt. in the third platoon. It was nice not getting shot at every time you took off on a mission. We had a Captain for a platoon leader that I would not wish on my enemies. We went on a long maneuver, "Air Assault II" for two months around a couple of Southern states and he got us lost one day and landed near a dairy. The cows did not give any milk for a week or more. That cost the Army a bundle. I showed him where we were on the map and he not believing, sent me up a hill to look for a road sign. The 82nd Airborne was the aggressor on the exercise and two of them, a Sergeant Major and his driver no less, saw us land and came creeping over the hill to see what was going on. I was captured by the enemy and taken to their HQ. It turns out that aviation personnel were off limits because they were needed to maintain the aircraft.

Later that night two MP's took me out on a lonely road and let me use their radio to contact my unit so I could get picked up. My CO wanted to know why and how I was captured. After telling him I showed him on the map the location of an artillery position that they took me to.

He called Battalion and our Battalion Commander got hold of an infantry battalion and we assaulted that artillery position at first light. This was not in the game plan for the war games but it came off so well that we had to stage it again for some congressmen/women and Army brass. You see the unit was stretched out along a ridge and they had not a clue that they were going to be attacked. They were completely and totally unprepared and had no guards out. We had come up a valley and then hopped up landing on the road right next to their camp and captured the whole lot without firing a shot. The GIs were standing around in their underwear brushing their teeth and shaving. That was one of the highlights of that field trip. They were literally caught with their pants down. Wish I had had the foresight to photograph the incident. The embarrassment and shock on their faces made the whole field trip worth it.

One of the farms that our battalion camped on during that an old farmer owned exercise that only let his children wear shoes for weddings, funerals and if it snowed. He discouraged them from going to school past the 6th grade. The farm had trees up to 10 feet high growing all over where there should have been crops and the Army had to pay for the crops that he could not grow during our stay on his farm and also we had to pay for any trees that got damaged. We knew he had not put in a crop for more than ten years.

Another and not so good was the way our troops were treated in some of the towns. Not having a shower facility some of us went into a small town and asked the sheriff if we could us the shower in the jail. "Certainly" he said. But when we came back to shower he said, and I quote, "You ain't bringing no niggers in here to bathe, they can go out back and use the hose."

One of our sergeants wanted to shoot the sheriff, but instead we got back in the truck and went back to camp. Another time we stopped at a truck stop for something and we noticed a black man standing in a mud puddle on the side of the store. When asked what he was doing there he said, "If I want to be waited on I have to stand here at this window until someone inside chooses to recognizes me, opens the window and asks what I want." The store kept the mud puddle filed with water all the time. Having grown up in a small mountain community in California this type of behavior was, and still is shocking.

Aside from our normal training, a mission came down to support the 82nd Airborne and a Marine task force in the Dominican Republic. Another battalion begged the mission away from us. I had been on all night duty as Charge of Quarters in our company office when our Battalion Commander called and said don't go anywhere and don't let any of your crew chiefs sign out on pass as he was coming in with my CO. The other battalion after begging for the mission could not make up a complete company of flyable aircraft, 20 helicopters. That Battalion was assigned the same number of lift helicopters that we were and could not make up one company of flyable? I had to wake up my platoon, four other crew chiefs, plus two more and our gunners to support the 229th. I opened the arms room and issued their weapons. Our CO told us what we needed to take on the mission and so with our gear packed and our toolboxes we were off, ending up taking seven of our helicopters to Dominica for a couple months.

We flew down to Florida and departed from there and as we flew off the ship, the LPH-5, we were issued three loose rounds for our weapons. You tell me what you can do with three loose rounds in an M-60 machine gun or an M-14 rifle. The battalion commander of the 229th was not the swiftest dude on the block. I did some trading with the infantry for more ammo. The Company Commander that we were supporting on the mission took it from us. My senior pilot, an American Indian that looked the part went in and chewed out the Company Executive Officer and pulled rank on him taking his job. He carried a big bad-assed looking knife and I think he may have offered to scalp that gentleman. Out I went trading again and got more ammo for my guys. It was a strange little war, we were not allowed to fire even when fired at and that was most every day. Most of the helicopters had been hit except for mine. My guardian angel was still at work.

The engineers had cut down all the trees in the back yard of the old Presidential Palace and the 18th Airborn Corps had taken it over as their HQ. We could land three helicopters there, one at a time and tying down the main rotor blades before the next one could come in. On a roof top nearby was an individual that took shots at all of us going and coming from that place. As usual we were not allowed to return fire. Even seeing the perpetrator we were not allowed to shoot him. Initially we were stationed at the local airfield and we discovered that the Air Force had real food for their troops and not just cold C Rations, but steak and fresh eggs for breakfast. We were a small group, so they did not mind us eating with them for the small allowance that we were paid for rations, about a dollar a meal.

On one trip along the coast we landed in a fishing village that was desperate for food, as the U.S. had stopped all shipping to the island. We went back to base and filled up both choppers with C Rations and upon returning to the village the pilots issued the rations to individual members of the village to keep the headmen there from selling them to the people. For whatever reason, we also visited a Banana plantation owned by an American. He was bellyaching because he could not ship his bananas to the states. The local villagers all came out to see our Hueys parked on the road by the plantation. Surprisingly most of them spoke enough English to communicate very well. It seemed that most of them wanted the conflict over so they could get back to their normal lives.

Along with supporting the ground troops with supplies we airlifted 105's in two pieces. They hadn't developed the lightweight guns that we used in Vietnam yet. Tried lifting a jeep and ended up dropping it on a stonewall to keep from crashing, destroyed the jeep. On a couple occasions we delivered footlockers full of money to the outlying towns so the police and other government workers could get paid. Guess who printed and supplied the money?

The U.S. of course. Special Forces were in control of a radio transmitter and we had to take a repairman and his tools to a mountaintop. Almost crashed there, the mountain was too high for us to maneuver on. The only thing that identified the Special Forces soldiers as American's were the 45's in their holsters.

Interesting thing about Dominica, we had to find a gas station that we could land at and get road maps to find our way around. It seems that we went down there and nobody had aviation or tactical maps of that Island country.

One of our crewmembers was a civilian commercial rated pilot. He met some Dominican fighter pilots and flew a mission with them, strafing a radio station controlled by the rebels and another rebel stronghold. He almost got a court martial for that. He said it was worth it. I believe they were flying some old P-51's. About that time we got moved from the airfield to an old polo field near a very nice hotel. The Battalion Commander did not like that we were eating with the Air Force. We all used the barbershop in the hotel and then got into trouble from the Battalion Commander for doing it. Go figure. We were trying to keep our military appearances up and got in trouble for it. There were 'many' other 'benefits' from being on the polo field. We were soon moved into an abandoned brewery so the powers that be could keep an eye on us. Sleeping on the floor of a Huey was far better than an old rat infested brewery even if they provided cots and mosquito nets. It still smelled like rat droppings.

While we were there the Army decided that my flight physical had expired and they sent me off to P. R. for a new one. I flew on an old 'C-119' with the Air National Guard. The pilots in a preflight briefing told us that when we went down, not if we went down, that we should keep our boots on because of the sharks. The C-119 has the glide characteristics of a lead balloon in a vacuum. On arriving at the air base in P. R, I was only given half a physical. Waste of time.

Back in Dominica, one day while walking in downtown Santo Domingo we saw a bunch of men drinking what looked like short long neck bottles of beer at a sidewalk bar near the American Embassy. No, it was short long neck bottles of dark rum, good stuff. I bought a dozen or so and wrapped them in red rags and hid them under the sound proofing, in the frame of my Huey. I left there a week or so before my ship came home. They were still there when my chopper returned and I had no trouble selling a few to the guys that had traveled with me. While in the Dominica we visited the American Embassy store a few times to shop for gifts to take home. I bought a pearl necklace and some earrings. Gave the Pearl necklace to my mom, still have the pearl earrings. Need to have the posts changed for my wife. After we got home from that trip I did such a good job of getting two of my crew chiefs to reenlist that I took six more myself and they only enlisted for three each. After a reenlistment leave back in California and returning to Benning the division was re-designated as 1st Cav. Div. Air Mobile and we deployed on the USS Boxer from Florida to Qui Nhon harbor and then to An-Khe South Vietnam. While I was home on leave I got engaged to a girl that had been the cook at the pack station I worked at in 1961. She used to write almost every day for the first few months and then the letters tapered off and stopped. Didn't even have the courtesy to write a 'Dear John' letter. Just as well, have not seen her since that leave, no loss.

Chapter Five: Co C 227th ASLT Hel. Bn, 1ST Air Cav Div, An Khe

All the way across the Atlantic, through the Mediterranean, the Suez Canal, the Red Sea just to get into the pacific and the South China Sea. Go figure! We could have been taken by train, across the U.S., then been put on ships in California and be there in two weeks, instead of a month at sea. On the way we refueled without slowing down. The ship announced that was a first for them. Going through the Canal we were told not to wear any Division patches so the pilots guiding us through wouldn't know who we were. By then every paper in most of the free world and probably the USSR had already announced that part of the First Cav. Div. was on the move and that we were on the USS Boxer.

The Boxer was probably the last carrier commissioned at the end of WWII. It sailed for the Sea of Japan with civilians still aboard and made it in time for the surrender and got credit for being in the war. She served proudly in the Korean War and had been converted to a Landing Platform Helicopter (LPH) that usually carried a Marine task force and all their helicopters. A friend in Sutter Creek, CA in 2006, has told me that her son was stationed aboard the new USS Boxer.

In the Mediterranean, approaching the Suez Canal with many ships in line waiting to get through the canal.

After landing in Qui Nhon Harbor I convinced my Co to spend the night there and fly to An Khe the next morning. I had ulterior motives to see if I could find any of my Vietnamese friends that I knew from before. Only found one that owned a restaurant in town. After warning some of the younger pilots what the effect of the local beer would have on them after not drinking for a month they went into town and got plastered. Some M.P's found them trying to get back on base; crawling in the early hours the next morning and feeling sorry for them brought them to our commander. He just laughed and told them that they had been punished enough for their actions. That morning we flew to An Khe with some of the crew chiefs flying in the co-pilot seats because of the pilot's hangovers. That night in the town was good respite after a month at sea. There was an advantage to having been there before and knowing some of the area and the people.

On to An Khe and the "Golf Course" as the area became known. It was unused rice paddies, each one just the right size to park a helicopter on. There were also many burial sites there that the Engineers had to move with the help and support of the local Vietnamese government. We set up our camp just over the hill from the Golf Course in neat rows of Pup Tents. The only large tent with us at that time was the Mess Kitchen tent and a small tent for the orderly room, Company HQ. Most all of our gear was sent to a different location as the Cav. was not going to all be together in one place. The orders changed and we all landed in the same location without much of anything. After arriving there, the first letter I received from my folks told me that a new car purchased in Georgia had not arrived yet in California. I had bought a new Chevy-II and took it to the Southern Railroad depot in town to ship home. A little over a month later it was not there yet. My dad went to Southern Pacific in Fresno and raised hell with the freight agent. My dad having been a Marine, a muleskinner and horse wrangler in his youth, had the verbal skills to peel the paint off your house. They apologized and said they had no idea it was coming and Southern Pacific called Georgia and raised unmitigated hell with them. The car was in Fresno a week later with over a thousand miles more than when I had left it with them and an apology. Southern Railroad personnel had been using my car for their personal use. Fortunately, everything that I packed in the trunk was still there.

In Qui Nhon there was row after row of brand new vehicles setting by the runway with full tanks of fuel and all their logbooks on the seats. After our only ¾ ton truck was wrecked we went back down there and "acquired" one of the new trucks and had our gunners drive it back to the base and did some Number switching on it. Never did get caught. Along with the truck, we found pallets of fruit juice, uniforms, boots and other goodies on the runway. Couldn't take the truck home empty, so fill it up we did and the helicopters also.

The beginning of the rainy season in a pup tent shared with another soldier was not ideal way to live. I could feel for the infantry in the field often not even having their tents to get into. After digging a moat around our tents, we dug another one in the back, set a pallet on it and wrapped our duffle bags in extra ponchos on the pallets to keep everything dry. It was not to long before our equipment caught up with us and we had our GP Medium tents. Each was large enough for 8-10 men to live comfortably and we put them on wooden or concrete floors and with folding cots and mosquito nets it became comfortable.

'PUP Tent City' Fun in the rainy season!

GP-Medium with floors. "Home Sweet Home."

Early in our stay at An Khe many of the troops would ask me to help them in town bargain for gifts to send home to their families. It was always a pleasure to get down scratching numbers in the dirt or on a scrap of paper until a reasonable price could be worked out and the Vietnamese vendors and shop owners enjoyed the exchange also, offering better deals the next time they saw you in town. Very few American GI's took the time to play the game and barter for their goods. The local vendors respected anyone that did. A Khe was a really small town with mostly dirt streets. The main street was paved and about a hundred yards or so long. With one bar, one questionable restaurant and many small shops. The open-air market was the place to shop for gifts. It wasn't long however that a dozen or more new bars and other things opened to take care of the new GI's in town.

The main street.

Open Air market.

The places of 'ill repute' opened just out of town in a field. Not a good place to spend much time. One young Jewish lad in our company had not known a woman before we shipped out, and this concerned us. He went to that area outside of town and fell in love with the first prostitute he met. Wanted to marry her and take her home. We had a hard time convincing him that was not a good idea. What finally did was the knowledge that his family would have disowned him. I think his father was a Rabbi. I'm not saying that the woman was so bad. It's just that it would have probably ruined both of their lives and disgraced his family.

If I had a day off, it was advisable to get off base or take a chance at being sent out on a mission. The one bar in 'old town' had a very sturdy shelf under the bar that that was about a foot and a half wide where the owner would let me sack out for a few hours. Brought my own pillow and left it for future use. While I lay there sleeping, G.I's would come and go, not knowing that one of their own was sleeping under the bar. Remembering my dad's advice of getting to know the people sure made life a whole lot easier.

I remember one African American gunner who fell asleep shortly after we arrived in Vietnam on guard, crawling inside one of the ships during a rainy and miserable night on the Golf Course. He was court marshaled and sent to a Marine stockade in Okinawa for punishment. A couple months later he was let out for good behavior. What soldier wanted to deal with the Marine Corps Stockade? He requested to return to our unit and begged forgiveness from the Commander, asking to return to his former status as a gunner. Permission was granted and he became one of the best men we had; a model soldier always clean shaven and neat in appearance keeping the rest of the gunners in line and letting them know how it was to be a prisoner in a Marine stockade.

For the first few months most of our missions were relatively uneventful. One day my ship was flying on the company commander's wing delivering supplies to some-one, can't remember who. We received a distress call from a Special Forces team in the Plei Me area. They were being overrun by enemy forces and had pulled all the survivors into the command bunker. They were requesting direct fire on their position.

Not being too far away and having already complet-ed our mission, we responded. The VC were throwing dead and wounded on the wire around the camp to use as bridges to get into the inner area. They were coming in waves at the camp. As we flew in a tight circle above them it was not necessary to even aim at the enemy. You just pointed your weapon down and held the trigger until it was empty. A call came from the bunker to evacuate the wounded and the medical evacuation helicopters refused to go in under fire. My CO went in; landing on the bunker's roof and his crew chief had to pound on the door and convince them that we were Americans there to help. After a brief pause in the attack he was able to take on as many wounded as he could carry and took off. The battle resumed as some gun ships came in and relieved us. We were pretty much out of ammo as we returned to the base in Pleiku with the wounded.

Once again, with a battle raging less than 100 feet below us my helicopter was not hit. I believe that a relief column of ARVN's eventually reached them and the Enemy having suffered heavy losses pulled back. A couple months later we were there again and waiting outside the wire for our passengers to come back out; we were kicking around in the brush that was starting to re-grow. One of our pilots kicked under a bush and out came a skull left from that earlier fight. By now pretty well dried up. That battle was the beginning of the story by Gen Harold Moore and Joseph Galloway "We Were Soldiers Once and Young." Thank you Mel Gibson for staring in the movie. As I recall it was the year of the horse in the Chinese calendar and the NVA thought their horse was stronger than ours, the horse on the 1st Cav. Patch.

The movie has a little Hollywood in it but by and large it is the most accurately portrayal of the conditions and what happened that I have ever seen. Even now I cannot read the book or see the movie without tears in my eyes. Moore's wife Julie was the driving force that got the Army to re-evaluate the way they were notifying survivors of their lost soldiers. It was a sad day a year or so ago that I received notification of her death. Julie and Harold were so very devoted to each other, their children and their faith in God.

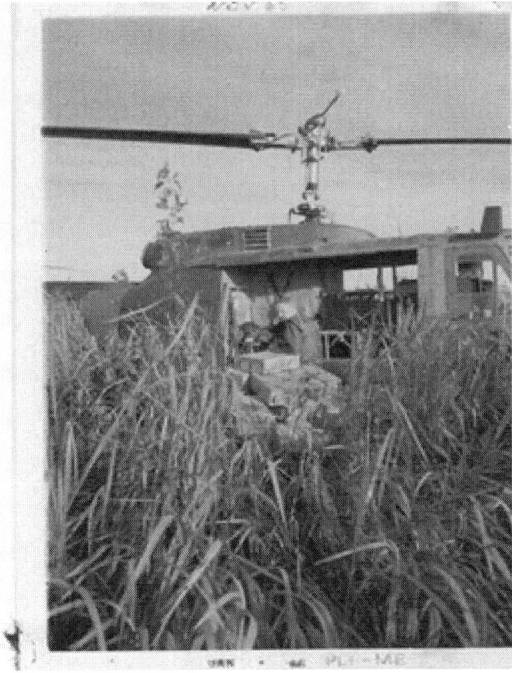

Elephant Grass. Sometimes gets more than 10 feet high.

The relief Battalion had to fight in this stuff against an unseen enemy.

My daughter took Sandy, myself, and our two grandsons to see the movie in Redding, California, shortly after it came out. I had taken two of my 1st Cav. Patches, one for each of my grandsons and after the movie as I handed them to the boys I couldn't, talk only cry at the memories that came out. I was a little embarrassed but Sandy told me it was ok to show my feelings.

Back to the Pleiku area, from that point on it got pretty hairy for most of November. We were in Pleiku, but I never got to see if any of my old friends were still in town. When Maj. Crandall got the order to support LTC Moore with only 16 ships we knew we would be in there helping out again. For some reason they seldom could complete a mission without help. The major was an excellent pilot and a brave leader and he justly deserves the Medal of Honor bestowed on him these many years later for his heroic actions in supporting LTC Moore's battalion.

For our part we filled in where needed hauling troops, ammo and taking out wounded. If you want more details read the book. You won't find the 227th in the book but we were always there to fill in where needed and help pull them out of trouble.

After the battle LTC Moore evacuated all, 100%, of his people, as he had promised, out of the valley and was relieved by another battalion that marched into the valley. Most of us never understood why they then had to march to another location to be evacuated. It served no purpose for those other soldiers to be left in X-Ray, but they were, and as they got strung out in another valley they were almost decimated after most of their officers and many senior NCO's went to the head of the column to see a captured NVA. This also is very well written about in the book by LTC Moore and Joe Galloway--enough said. Once again we got involved with rations, ammo, replacements and evacuations. I had never seen so many Americans shot in such a short time. I'm only an enlisted man and didn't know their commander but they didn't seem to have the leadership that Moore gave to his Battalion and that seemed to be at least part of the cause if not all of it.

While still in Pleiku, we flew to the Catecka plantation with some Army Brass. The owners had been put out of their house and moved into the servant's quarters so the Army could use the house as a headquarters. They also had their small plane taken from them for fear that they might be supporting the enemy? That plane was their only transportation out of the area for shopping and visiting friends. The roads were not safe for travel. I did not get to see the family that owned it. If I could have, what could I say to them? We were friends a couple years ago and now we kick you out of your home and take your transportation from you!

After over 40 years I am not sure of the sequence of some of the events that took place with the 1st Cav. but as I sit here things keep coming into my thoughts. One day that was very long and into the night we got word that some infantry had walked into an ambush and were out of range of any of our artillery. My platoon leader got us back in our ships sometime after dark and finding a green spot on the map close enough to the cut off Unit for artillery support; we assaulted the "green spot" on the map and with no knowledge of the area and with a bunch of Rangers. They checked the area and set out lights so we could bring infantry in to secure the area and the heavy helicopters could bring in 105's and their supporting equipment to help the trapped unit. They were hit so hard that one young 2nd Lt. was the only officer left standing and he had just got in country. He had about a squad of men left that could fight. After setting a perimeter and setting out his Claymore Mines, he sent out scouts to look for the enemy. Finding none he called for evacuation. The Chinook from the 228th that came in to evacuate them came to a hover and was shot down.

The NVA had been hiding, waiting for just such a thing to happen. The pilots, crew chief and gunner grabbed their weapons and hit the dirt. The young Lt. was grateful for a superior officer to take command. The senior pilot said "No you have the situation under control; just tell us where you want us." The Chinook had two M-60 machine guns and plenty of ammo. The pilots may have M-16's or shotguns and that along with their crew chief and gunner helped save the survivors of that unit as well as their own selves. The artillery fire must have discouraged the Enemy once we got them in place. In the morning we flew in there to pick up the survivors.

They had loosely covered the dead NVA because of the stench of dead bodies in the heat and humidity and were still maintaining tight security in their small area. The downed Chinook had to be sling loaded back to An Khe for repair. I never remembered who that Lt. was but I hope he was recognized for his bravery and stamina against a far superior force of maybe an entire battalion of NVA. If he did not get the Congressional Medal of Honor for that action, the Army did him an injustice. Even if it was the year of the horse on the Chinese calendar, that entire year the 1st Cav. Horse had out shown the enemy's horse at most every turn.

In maintaining a UH-1 helicopter there are inspections at 25-hour intervals and a major inspection at 100 hours. There were times that we would not shut down for up to 20 hours or more, just refuel and go again. During those times we would do the physical "look see" inspections at refueling stops and log them into the logbooks. Only if an oil sample from the engine or transmission were due would we ask the pilot to shut down for a few minutes. Most of our gunners were familiar enough and trustworthy to assist with some of the inspection criteria. If we had trained them, they could take oil samples while we lubed the rotating controls on the rotor system. Most of our gunners became very close and an important part of the team and not just as a gunner.

There was one sad note near our camp. Someone in command decided that the wild elephants could be captured by the enemy and used to transport heavy artillery against us. Some knucklehead at division sent down the order to destroy any wild elephants in our area. Don't think he had a clue as to how we were to kill an elephant with the weapons that we were issued. We were coming back from a mission when we received a call for help. Another slick ship was trying to kill an elephant with their M-60 machine gun. The top of the poor animals head was beginning to look like hamburger when my pilot called for a gun ship. They were able to put it out of its misery with a pair of rockets to the head. Someone got in a lot of trouble for that order and it was rescinded. Another sad note; we had a dog named PFC Charlie Rider that adopted our company, he was hit by a vehicle and seriously injured. The Division surgeon got together a surgical team and tried to save him. Unfortunately they couldn't, his injuries were too severe, but they tried. PFC Charlie Rider was missed by all and had a proper military funeral.

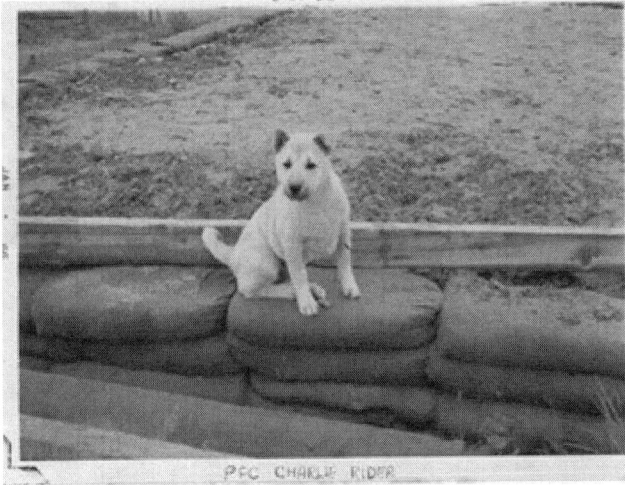

He wasn't big, but everyone loved 'Charlie.'

The Tiger Division from Korea, all volunteer, had come to Viet Nam and engineer units from that division cleared an area for us by the coast and provided most of our outer perimeter security. The Korean engineers that were clearing the area and setting up the perimeter defenses came under attack from the V.C. Defending their positions they all died, pulling their weapons under their bodies to deny the enemy, and taking so many with them that the V.C. made it a point to not engage them in the future unless they felt they had the advantage. We arrived on scene that day to late to be of any help.

My ship had been acting funny with a strange vibration and I grounded it only to be overridden by the maintenance personnel that were with us in that forward position. I grounded it again and demanded that one of the Tech Inspectors and a maintenance officer from An Khe come and check it out. Good move. The center mount under the transmission was the only one not broken or at least cracked. You must understand that I have had this aircraft for almost two years and it is the one I took to the Dominican Republic. It was also one of the first "D" models that came out of the Bell factory and because the "B" model rotor system that was installed on them did not provide enough lift, a 2 foot section was added between the fuselage and the tail section so that longer blade could be installed. The helicopter wasn't new when I got her at Benning. Waiting for parts I was stuck on the ground helping with the other ships for a while.

One day we were told that an attack was imminent and I was assigned all of the extra personnel to form an inner security defense near where the helicopters were parked. I picked the best machine gunner that I knew and assigned him to be with me in the center foxhole. My position extended for about 100 feet on both side of my hole, that way I could be in communications with all my guys. Digging a foxhole is somewhat of an art. A two man position should be somewhat in the shape of a "U" with a sloping floor and a hole dug as deep as you can reach in the lowest spot to kick grenades into. My partner saw me digging another hole in the upper corner of my position and wanted to know why? I just grinned and dug a hole in the side for a roll of toilet paper. He groaned remembering that I had a bad case of dysentery.

Why is this important? It eventually got me back to An Khe before my ship was repaired. Shortly after leaving they were attacked and one of our crew chiefs that had been sleeping on top of his helicopter in a sleeping bag was killed. He rolled off the helicopter and no one noticed that he had been shot until he didn't show for breakfast in the morning. My replacement in my foxhole did an excellent job and helped prevent the NVA from overrunning our position.

Back at our base camp a friend known, as "Slack" needed some flight time for pay. I let him take my ship out and when he came back, he had forgotten to clear his weapon as they flew over the perimeter of the base. On checking it he fired a round through one of my fuel cells just after I stepped back from putting my gear back in the chopper. Slack was an African American and his face turned ashen when it happened. My Commander asked me what I thought should be his punishment and I told him, no further punishment needed, if you had seen the look in his face after he almost shot me. It took a week or so to get a new fuel cell and get my ship back in order again.

Close to Christmas time everyone was excited to have Bob Hope and his traveling USO show at the base. Unfortunately for my platoon, we were selected to stand by with a platoon of infantry in case of trouble. We could hear him and the 'round eyed' women singing but could not see them. So much for our entertainment. That was the only time that Bob Hope came to an area that I was stationed at.

War is hell; well it can be but even in war if you put the best spin on things it can be rewarding. Being on my second tour over there and fortunately in aviation, I was able to take advantage of the time and location, visiting places that I remembered from my first trip like the city of Dalate up in the mountains. Dalate is or was the location of the Vietnamese military academy, university and a former

French resort town of sorts with pine forests, trout in the streams, deer in the woods and also tigers and elephants. In the center of town was a lake with paddleboats not unlike the ones you might see here in this country. I remember seeing a young couple on one of them obviously unaffected by the war going on a short distance away. They seemed to be enjoying each other's company and appeared to have a picnic basket with them. On one trip up there I meet a very attractive young woman that owned a restaurant and we were staying in a very nice almost new hotel with hot showers and flush toilets. That may not sound like much to you; just try living without them for a year. This hotel was right across from the town market place and I was awakened early in the morning to see a tiger eating what appeared to be half a pig outside of the market. The market was built on the side of a hill and was two or three stories high. Having no refrigeration they would throw out meat after a couple days and the local dogs and an occasional tiger would clean up for them. The city had many beautiful churches, hotels and restaurants and even the 'peasant' lived in wood framed houses with glass windows and not the rice straw and mud homes that we saw in the Delta.

A few months earlier everyone was getting tired of canned rations and I suggested that we go to Dalate to purchase fresh veggies and fruit. The Commander thought that was a good idea and after investigating the location we took two ships and brought back fresh strawberries and lots of cabbage that was in season at the time. After observing the town he set up a three-day pass rotation to Dalate. That was a welcome relief from our normal routine. We even flew back-corrugated roofing for the hooch's that we were building to replace our rotting tents. Along with the fresh produce that we now had in the mess hall, one of our cooks that had been a chef in New York asked everyone to send home for pie plates and Ritz crackers. He made the best imitation apple pies that could be made. This cook could make powdered eggs and spam into a gourmet delight. If the food didn't look good and taste good he would refuse to serve it. Drove the mess Sgt. crazy. Our dinning hall was the first large building that we constructed and the center of Company activities. One of our Staff Sergeants after a couple of beers would loose some of his inhabitations and bring out his guitar. That man could play some of the best classical guitar you have ever heard. When he started to play, people from the surrounding companies would crowd in to listen, even standing out side in the rain by the screened windows.

While on the subject of time off; a friend and I took a seven day in country leave to Dalate and started hitching rides, first to Qui Nhon and found a really good restaurant. Ordering lobster the waiter told us, "One lobster, two G.I." One lobster tail was almost more than we could eat. Our dinner included melted Danish butter for the lobster, salad, strawberries for desert and a good Portuguese wine and wonderful French bread made from rice flour. The next day we caught a ride to Saigon. More good food and as I was showing him the parts of the city that I was familiar with, we were walking down a street that we used to call the street of flowers for all the flower vendors in the center divide, when a shop keeper came out from his jewelry store and greeted us. He remembered that I had been in before and had spent over a week getting him to come down a couple dollars on the price of some jewelry. We had tea and lunch with him and his family at the store. The next day we caught a flight with some Australian Airmen going to Dalate. The Australians are a very laid back bunch of guys; just don't eat with them unless you like Mutton. One of our pass rotation flights were going back to An Khe at the end of our leave. Forgot to mention, in Dalate I had the best cream of Asparagus soup that I ever have had. Just before serving it, a beaten egg is stirred in gently, almost getting cooked in the process. By now you must be aware that I like to try different foods wherever I go. I'm still the same only now I really enjoy cooking.

One of my good friends from Fort Benning was Sp/5 Melvin S; he also volunteered to go to Viet Nam in hopes of a promotion before he retired. A mortar platoon had been left on a mountaintop on our outer perimeter without food or water and it was during the rainy season. Their Commander thought we could always take what they needed to them. Being locked in the clouds for a few days, my commander decided that we should try to get to them. Hot food and water was loaded in Melvin's helicopter and they tried to hover up the side of the mountain just close enough to see the trees and far enough away to miss them. A tree was growing nearly horizontally out from the mountain and they flew into it. The mortar team on the hill could hear them coming, hear them hit the tree and hear their screams as they died but could see nothing for the clouds. I still remember how devoted to each other Melvin, his wife and children were and have often wondered how she has managed after his senseless death.

All because some commander didn't think his men should have enough rations to carry them for a few days if he couldn't get to them. Melvin and his wife had been married right out of high school and had the classic made in heaven relationship. Our CO tried to get him a promotion after he died. Unfortunately, division turned it down, saying they could not postdate the promotion. When the traveling Vietnam memorial wall came to Monterey, CA many years later, I found his name and could not help but shed a bunch of tears thinking of his family and their loss. Melvin had remained true to his family, never going out 'on the town.'

One day, assaulting a village next to a river with several flights of Huey's behind us we had to take off flying across the river and up a steep mountain. As we flew up the mountain barely above the trees I spotted a column of NVA coming down a ravine and out of instinct, popped a white smoke and opened fire getting two or three of them from about 30-50 feet away. Once again I was told to stop firing, they might be friendly; not with a red star on the helmets and AK-47's in their hands. The rest of the flight behind us diverted, and the infantry being alerted did not walk into the trap that was being laid for them on the edge of the river. Division had dropped leaflets earlier over the village telling them that we were coming, so all the civilians had left town and the NVA had time to get ready across the river for our arrival. Go figure, nothing like telling the enemy we are coming. As I recall, our troops took a heavy toll on the enemy that day.

Taking a hot breakfast to an artillery position on a mountaintop, as we landed we noticed everyone was gathered around a large tree on their perimeter. At night the artillerymen usually put a round in the tube that contained steel darts and lowered the tube horizontally with the ground. Upon hearing sounds in the night the crew obtained permission and fired. In the morning nailed to the tree was a VC sapper in a crouched position with his satchel charge still clutched in his hand very dead, the steel darts had once again done their job.

I even got to take a 7 day R&R in Singapore, "The jewel of the orient." A city so clean that you can drink the tap water in most of the city, just don't spit on the streets and now, do not even think about chewing gum. Singapore is a City State and was influenced greatly by the British in a very positive way, Orchid gardens, museums, good food and wonderful people.

Walking into a variety store the first morning I was there, a crisp, very British voice asked, "May I help you sir?" After being in Vietnam for so long I was embarrassed to see a young Chinese woman, sales clerk, standing there looking at me.

The next night with the hotel social director as my guide, a young Chinese woman, we went to the American Club at the American Embassy. There we met an American engineer and his Eurasian secretary. Often wondered if they ever got married. They had been together for several years and neither was married. Around midnight she would say; "now honey you know we have to be at work in the morning," and his reply usually was "Do we have to." He was in charge of a project to deepen and enlarge the harbor. Singapore is a free port and uses local dollars and cents as its currency. Their dollars were only worth about thirty three cents American.

What a shock to buy a pack of cigarettes and pay a dollar for them. We also met the Navel and Air Attaché to Singapore from Canada, a Canadian Navel Captain, and a very interesting gentleman. Most of the clubs in the city required a tie and jacket that they would loan you. After an evening out, my friend would open the kitchen to make us a snack and then we would swim in the hotel pool. She took me to one club that the officer in charge of the R&R program liked to hang out in. He had been trying to get a date with her for the three months he was there and without any luck at dating the young woman. I avoided him for the rest of my time in Singapore. I was having an English wool suit made at the "Raffles Hotel" by "Chan He Tailors" in the hotel lobby. The Singapore Lions club was meeting and I wanted to attend. That would have been fun sending documentation to the club that my father and I belonged to in Oakhurst. Unfortunately the tailor wasn't happy with the way my suit fit and I missed the meeting.

That evening I was talking with the bartender at the hotel we were staying in and he decided I should meet his sister. She was a very attractive Eurasian woman in her mid to late twenties and was the chief telephone operator for the Singapore exchange. We had dinner at the hotel the next evening after visiting the Orchid gardens and the House of Jade. It is owned by a large oil company and has one of the best-carved jade and quartz collections in the world. Unfortunately the next morning was my return flight to Viet Nam and the war.

LTC Cranford, our battalion commander was getting ready to rotate back to the states so it was decided that he needed an appropriate going away gift. With another ship that was carrying the Battalion Sgt Maj we headed for a better PX than we had in An Khe. Crossing a small village on the coast the other helicopter went down with mechanical problems a short distance from the village. Flying low to find out what they needed we took fire. When I returned fire my pilot yelled. " Don't shoot you will start something."

By this time I was getting used to being told that. The village began shooting at the downed chopper with our SGM on board. We went up to about 10,000 feet to radio back for the parts needed to repair the ship and to call for assistance. A flight of Huey gun ships from Qui Nhon was returning to their base without having fired a shot so they came by and pretty well obliterated everything on the ground. After that a flight of fixed wing AD-6s that would have had to dump their bombs in the sea before landing came by and added to the mess. Next we received a radio call from a ship in the area that wanted to sail in and shell the village. We asked them not to as we had friendly's on the ground between them and the village. The Navy reluctantly sailed away. Once the other helicopter was repaired we went to Qui Nhon to refuel. My pilot was still saying that I started something so I took him around and showed him a bullet hole in the engine cowling and showed him that the engine temperature-sensing element had been shot off. He never even noticed that the temp gage on the instrument panel wasn't working.

About this time we had been there at least 10 months so I grabbed the first replacement crew chief that came into the company and turned my old tired Huey over to him and started working with a Capt. that was a contractor in the real world. He sent a 2-½ ton truck to an engineer storage yard on the highway to Qui Nhon with a Lt. and a bunch of enlisted men on a Sunday afternoon.

The gate guard let them in not knowing they were not authorized and they filled the truck with building materials from a list that the Captain had given to the Lt. on a clipboard. Also on the clipboard was a Mad Magazine that the Lt. was reading most of the time. So we started building wooden hooch's to replace our rotting tents. All of the concrete for the slab floors was mixed by hand in troughs that we made for that purpose. I made special Dutch doors for the Company Commander and the first Sgt's hooch's. The latch on the outsides were horse heads and on the insides, horses assess with the tail as an opening lever.

Unfortunately the crew chief that took over my Huey must not have been a very good mechanic. A few months after leaving Vietnam I heard that he had not been doing his daily pre-flights very well. They went out one day and the main rotor came off right after dropping off a bunch of grunts. The whole crew died and they found his body some time latter in a tree. This is only a guess as to his competence I have never talked to anyone directly involved in the accident. Rotor Heads do not just fall off! The pilots are also responsible to climb up and inspect the rotating controls before taking off. Who knows what went wrong?

There ended my second tour to Vietnam. After a 30-day leave at home and picking up the car that I had bought just before shipping out to Vietnam, it was off to my next assignment.

Chapter Six: Fort Carson, Co.

Thus begins of two years at Fort Carson, Colorado. At Ft. Carson in Company B 5th Avn. Bn. Later to become 283rd Avn. Co. the aviation maintenance personnel lived upstairs in the hanger in what was meant to be offices. There was a large restroom with showers so the mechanics lived there instead of the old barracks on main post. On arriving I was asked if I wanted one of the aircraft that were assigned to the company. My answer was once again; if the crew chiefs are doing a good job, don't mess with them. The unit I was assigned to at Carson was the only non-deplorable unit in the Army. Every one had just gotten back from Vietnam or was otherwise not deployable to the war. Not long after arriving we had a full field inspection in our rooms and the people that lived off post had their inspection on the hanger floor.

The Post Commander came in and with him was the Sergeant Major that was with my battalion in Vietnam. He sent the General on his way and we caught up old times. He was now the post Command Sergeant Major. During the inspection something happened and the fire sprinkler system in the hanger went off. Later someone said it was due to a sudden pressure deferential in the system. Just imagine, everything that you are issued all laid out on the floor with your rifle field stripped on top and it rains about 6 inches in 5 minutes? All their clothing, blankets, and sleeping bags had to be sent to the post laundry to be cleaned and dried. Everyone got a day off to get their gear back in order.

Shortly after that I looked up a friend that lived in Pueblo, Colorado that had served with me in Vietnam. He had a girlfriend there and her folks had a house for rent. It wasn't authorized for me to live off post, but I moved to Pueblo, 38 miles from Carson. The house was built in the 1800's and the kitchen and bath were added on the back of the house. The original kitchen was in a separate building so the house would not burn down with it. And the original outhouse, well that was still out in the back with a crescent cut into the door, enough said.

There was no garage, but a carriage house facing the alley. The whole house was furnished in antiques to include the old gas stove that had to be lit with a match and a "Hoosier Kitchen" cabinet with all the parts. Coffee grinder, flour sifter, salt and peppershaker and many more goodies. What I would give to have that kitchen cabinet now. The 'new bathroom' had a claw and ball tub and a pull chain to flush toilet. The kitchen and bath were add-ons to the back of the original house.

While in living Pueblo, one of the ladies that I dated had a cousin in the local police department. She would borrow her grandmother's Cadillac and we would meet her cousin at the lower end of Main Street. He would put on his light and siren in his patrol car and we would drag Main. He always had an excuse why the Cady beat his city Ford. While on the subject of the city police, at the entrance to the police station there was a large planter on each side of the main door with nothing in them.

Some enterprising soul planted some 'happy tobaccy' in the planters. When they started to grow, the police watered them without having a clue what they were. A few months later, a state narcotics officer came by and just about fell over when he saw the plants growing there. He pulled them out and confronted the chief of police about his Marijuana farm on the front steps.

At the Long Branch Saloon, owned by a Sicilian named Joe-Joe, one of the local police used to come in nightly for his 'Coffee' and usually brought a young woman with him, give her some 'Coffee' and then take her to his patrol car. This guy was a total idiot. On New Years Eve he was doing his usual and my partner took a package of lady fingers with the fuse wrapped around a Camel cigarette for a fuse, lifted his handcuffs and put the surprise in his handcuff case. We figured it wrong and they went off almost a minute after midnight. His feet were hooked in the rungs of the bar stool and he fell over backwards when he tried to stand up. Joe-Joe told us to send our girlfriend home before things got messy. The cop after following friend Jim into the restroom took us outside and had us spread eagled against the wall, checking pockets for firecrackers. Putting both hands into my front pants pockets, I backed up and his hands were stuck. Meanwhile, a patrol car with the Lt. and Sgt. on night shift pulled up with their lights off. They were laughing so hard, they almost cried at his antics. He found no firecrackers and for his foolishness as well as having an underage female in the bar, he ended up walking a beat in a residential area on the outskirts of town. So much for my association with the local police.

Pueblo had a city park with a good-sized pond that would freeze over in the winter thick enough to skate on. I had my folks send me my ice skates and with another young woman that I dated would go skating on the weekends. Never got real good, but had fun anyway. We went to the Broadmore Inn in Colorado Springs and watched the young woman ice skater practice that got the Gold Medal in the Olympics that year.

Not having a real job at the airfield, I had talked to the First Sergeant and he created a job for me. He had a desk put next to his and I became his assistant, taking care of all the duty rosters and anything else that he could think of. He was a fishing fanatic and if it wasn't lake or stream, it was ice fishing in the winter. He even had a part time job washing trucks at a truck stop in Colorado Springs. Sometimes we would not see him for days on end. I was glad to have a job and not just be hanging around the airfield with nothing to do. Occasionally on Fridays I would be sent to the COORs distributer in the city and pick a keg of beer for a company party on Saturday afternoon. And what was left would be served on Sunday in the mess hall after lunch. In the Army all the beer has to be no more than 3.2% alcohol. The distributer would take from his desk a red dot about two inches in diameter and stick it on the end of the keg and say, "You are now 3.2 beer."

We had one fellow that begged an emergency leave to his home in San Francisco and got it. About a week or so later we received a call from the MP's at the Presidio of San Francisco requesting that we send someone to pick him up and return him to Colorado. It seems that when he got home he held his sister in law at gunpoint, raping her and making his brother watch. He was captured shortly after by the S.F. Police and thrown in the slammer. For some reason at his preliminary hearing the civil authorities turned him over to the military under the condition that he never return to San Francisco. As was so often the case, the First Sergeant was gone, so I had to get special orders drawn up for two NCO's to go armed on a flight and pick him up. On arrival back at Carson he was complaining about the orderly room being too hot while I was preparing pretrial confinement orders. I let him outside with an armed guard. One of our cooks came running into the office yelling that the prisoner was running across the parade field with his guard in pursuit. Grabbing my revolver from my desk drawer I went after him loading as I ran. When I caught up with them, they were both on the ground scared and exhausted. The prisoner scared of being shot and the guard afraid to shoot him. The prisoner looked up at me and asked if I would shoot him. My reply was; "if I kill you I would be transferred from here and I don't want to leave, so don't run again." Handcuffing him to a jeep I took him to the Post Stockade. He was boasting about what he was going to do when he got there, so I told the officer in charge and he showed me a cage that was in the middle of a barracks room for problem prisoners. Because he had been charged in California with the felony we could only charge him with conduct unbecoming a soldier and give him a dishonorable discharge. The MP's took him to the gate and the local law took him to the bus terminal and watched him buy a ticket for San Francisco. They called us and we called SF and they were waiting for him when he got there. Obviously he wasn't the smartest person on the block.

I spent a lot of time exploring Southern and Western Colorado with my friends that lived in Pueblo. Much of the countryside was like stepping back into history. Even considered staying there and bought a house in Colorado Springs and designed a home to build in the southwestern mountains when I got out of the Army. I was dating a woman a couple years older than I was. She was Apache Indian and Castilian Spanish. Just don't make the mistake of calling her a Mexican.

One of my friends took me on a trip on a narrow road to an old country store in the Southwestern part of the state. The store was made of logs and the first couple layers had rotted away leaving the entrance door rather low. They had an old gas pump outside that you had to hand pump the gas into the glass upper section and then it poured by gravity to your gas tank. The owner said, "Why should I get one of those new fangled things that need electricity to work?" Some of the stock on the shelves had been there for almost a hundred years.

The first winter I was in Colorado was colder than I had ever known in the Sierra's. The next January everyone was praying for a blizzard and I thought they were crazy until the wind started to blow and everything got sand blasted. Against my better judgment I drove into Carson and when I arrived the wind took my car door and me almost off in the parking lot. By the time everything was secured in the company area and guards were posted at all doors I was allowed to go home. There was no paint on my car except under a chrome trim on the passenger side. All the glass and light lenses and chrome were also sandblasted. I had to drive with my head out the window to see the road.

The next day when the insurance adjuster arrived he couldn't believe that the insurance that I had taken in

Georgia had "0" deductible. They even paid to have the engine drained, flushed and re-filled with all fluids. All it cost me was five days car rental waiting for them to finish painting and replacing all the damage.

While at Carson a division of Colt Industries that made the fuel controls for the Huey came there to make a training film. It seemed that most of the fuel controls removed and shipped back to the states from Vietnam were not properly preserved and packaged for shipment, causing them to be discarded on arrival at the overhaul facility. These units cost $10,000.00 each and upon opening the sealed shipping cans, if there was any rust, they were un-repairable. The tolerances in the fuel controls were so fine that they were measured with a beam of light through a prism. It took us about three days to complete the film, starting on a cool but not too cold day and having to move inside the hanger to finish. It started snowing on the second day requiring us too start over. Got a nice letter from them thanking me for my effort? It was a nice diversion from the normal routine.

While at Carson we were assigned a mission to support a civilian weather missile program on Johnson Atoll in the Pacific. Loading a Huey in an Air Force cargo aircraft, 5 or 6 of us left for warmer climates. Johnson Atoll is a man-made Island with no females on it. On Johnson there were only military personnel and contracted civilians from Hawaii to do the cooking, cleaning, dishwashing etc. We spent a lot of time fishing and diving for corral, tough job, but some-one had to do it. The life expectancy of a wheeled vehicle over there was about 6 months. After that you couldn't find the spark plugs for the rust. An Air Force Colonel was the senior officer on the island that was controlled by the Navy.

Nobody seemed to know why he was there unless it was to keep him far away from every thing stateside. He had an Air Force flag up the flagpole and insisted that it stay. I designed a flag to represent the Navy, Army and the weather people that we were supporting.

On the island was an old sailor with no family that was allowed to stay in the Navy until he died. He had been a sail maker in his early days before the Second World War, so he made the design into a flag and the Navy ran it up the flagpole taking down the Air Force flag, removing the rope and greasing the pole with bilge grease; the nastiest grease in the world, salt water won't touch it. The Air Force Commander had a cow trying to find out who put that flag up and took his down. Every one played dumb and the cherry picker used to do the deed suddenly was not running. Down for parts that had to come from stateside, or so the Navy said.

Coming back from there, we arrived in Colorado and no one was there to help us get the helicopter off the cargo plane, a C-141. The pilots were getting angry, as they wanted to get back to Travis Air Base and their families before a mandatory rest period. Not only was our help not there, it was snowing and we only had light weight flight suits and deck shoes to wear. About a month later my company received a very nasty letter from the Air Force in Washington about the flag. Naturally we denied the whole thing, publicly anyway! Still wonder if someone was able to get and keep that special flag. I'm sorry that no one took a picture of it before we left.

On getting back to Carson we found that a loafer from the airfield that couldn't even pass his proficiency exam had been promoted to Staff Sergeant in our absence and the absence of our Company Commander. After apologizing they gave me the next promotion that came to Fort Carson. A couple months after my promotion I came down on orders to go back to Vietnam after attending Helicopter Maintenance Inspector School in Ft. Eustis VA or 'Fort Useless' as we preferred to call it. On arrival there I was called into the school commanders office, he wanted to know about my DFC. Not many enlisted men received that award, it primarily went to aviators. The school hours were from 6:00 in the evening until 2:00 in the morning for six weeks. After class you marched back to the barracks, took a cold shower and sweated yourself asleep. One of my classmates had his car there and we took advantage of having the days off and spent those visiting historical sites in the area. A couple years later I would run into him at Ft. Rucker working in a snack bar as a cook.

We graduated on the 5th of July 1968 and I flew back to Colorado to say goodbye to my friends, get my truck that I had traded my car on, pack up all my things and head for home and a short leave. The realtor that I had bought a house through managed the property for me and I left most of the furniture that I had acquired there. A partially furnished house was much easier to rent out, especially to young GI families. This time leaving the truck for the family to use, and heading back to Oakland Army terminal for the third time. That place is depressing; you can't process through that place in less than 5 days while waiting for orders and a flight out of Travis Air Base.

Chapter Seven: Final Trip to RVN

Arriving back in Vietnam in August of 1968, I was assigned to the 602nd TC Det. with the 187th Avn. Co. South of Saigon. The nearest town was the same one that in 1964 some of my pilots got the snake to get rid of the rats under the chapel in Saigon. Small world! Here I am a helicopter maintenance inspector on my first assignment after school. Much easier and safer job than flying most every day but I still got in a lot of flying on test flight, mostly after dark.

One of the first things I noticed was the officer's bunker. You have to picture this; they built a frame structure with siding, inside and out of 2 X material and poured sand into the spaces between. This was fine for stray bullets but when a mortar round or rocket landed to close and blew out a section of the 2 X material, the sand flowed out like water. Anything left of the siding would not even stop a rifle round. That was bad enough and then I saw the NCO bunker. Another joke!

They had taken engineer culvert halves and laid them end-to-end and covered them with sand bags. Here you have a half round tunnel about 3 feet high and 30 feet long, if a mortar or rocket landed at the end, every thing and every one inside would be destroyed. Designing a new bunker and turning in a material request to the supply section, they told me it would never happen. I said please try and about three weeks later all of the material arrived.

The bunker was built like a log cabin of 6X8 bridge timbers with a timber baffle inside the entrance and outside also. The roof was of steel runway paneling covered with three feet of sandbags and the outside walls had sandbags starting out at 6 feet and tapering up to 3 feet at the roof edge. We ran telephone wires in through a pipe and the company clerk would join us with a field phone during an alert. Two or three months later the officers got the message when they weren't allowed to use our bunker, accept for a few Warrant Officers. They constructed a bunker that looked on the inside like a ranch style house using the same methods as ours and used it also as the command and communications center, good move. The best part of the NCO bunker was that it was next to my hooch. At night I opened my flack vest and put all of my clothing, boots and revolver in the vest. In emergencies all I had to do was grab it up, run out the door and around the corner.

The VC and NVA had a habit of targeting the hospital that was right next to our company as well as the fuel and ammo locations. Not being the best shots we took a lot of it. There was a tower with a dual 40mm set up on top with back tracking radar on the Filipino side of the base. As soon as they fired a mortar or missile the radar would back track and fire within a few seconds. That could have been the reason for their inaccurate firing as they had to get out of the area fast to stay alive. That 40mm with high explosive rounds would pretty well saturate the area that they had been in.

Within that base and right next to our compound was a Filipino Army artillery unit that had fire missions almost daily. Often over our area and that was the cause for alarm. Most of the 105 rounds that they had were left over from Korea and WWII. It wasn't unusual to have air burst a hundred yards out of the tube. Whenever they had a fire mission over our area we automatically headed for the nearest bunker until they stopped firing. Special Services had built a real nice house for the girls assigned there and covered it with sandbags. It and the service club were on the Filipino side also. If you knew the Filipino Provost Marshal you could get through the gate after dark, otherwise daytime only.

We also had a NCO lounge rather than an NCO club. It was in the back half of the First Sergeants hooch and like the rest of our hooch's was well sand bagged. Everything in the lounge was a quarter and no money ever changed hands there. At the end of each month two NCO's would go around collecting and then go to the PX with their ration cards and re-stock the lounge. By having a lounge instead of a club, we did not have to pay the crooks in Saigon that ran all of the NCO clubs.

Late one afternoon a call came from our Battalion HQ for me to put together a recovery team for bunch of Huey's that had been shot down that day. They sent a Huey to pick us up, and we didn't know that the other companies in the battalion had pled ignorance of the sling procedures. The Huey's are down in rice paddies about 200 yards from the tree lines, all shot up and with blood on everything. These Huey's are sitting in about a foot and a half of water and I had to show my people how to rig them for sling loading.

The Air Force is flying overhead dropping flares to light the area for us. Battalion sent in Chinook's to sling them out and I told the young Lt. that was securing our area to radio the helicopters and remind them to key their microphones as they came in to discharge the static from their choppers. Assuming that they did and standing on the rotor head of the first to be lifted out I raised the nylon strap to hook into their cargo hook. The pilot did not do as directed and the static blew me off the helicopter and into the rice paddy! Almost was tempted to shoot the pilot. That fall from over ten feet and into the water was not one of the highlights of the evening; neither was getting sniped at from the tree line. After getting them all hooked up and gone I asked the pilot of the Huey that picked us up to be sure that our ropes and sling straps got back to us at our base. His comment was; you are not going home tonight, there are more for you to sling out tomorrow. We ended up at someone else's base, cold, wet and without having anything that evening to eat or dry cloths. Their Mess Sgt. Was indignant about getting us something to eat and the Supply Sgt didn't want to give us blankets to sleep under. The next morning, still wet and unshaven we got chewed out by the Battalion Commander for our appearance in the chow line, at which time I reminded him of his heritage ending the comment with; "Sir" and after breakfast we went out again to recover more helicopters.

This time there was more aircraft than we had gear to sling properly. Doing our best, one of the Huey's being slung started swinging back and forth under the Chinook almost causing them to lose control.

I told the pilot of our Huey to radio them to drop their load and have one of the Chase Huey's go down and put white phosphorus grenades over the fuel cells so the enemy could not recovery any parts. It burned very nicely on crashing to the ground, did not need the grenades. The battalion Commander remembered me and stopped a recommendation for an award for my recovery crew and me. He also rescinded a promotion that I received after going home on emergency leave giving it instead to his driver that was a draftee with less than a year and a half in the Army; a draftee making Sergeant First Class?

The Maintenance Officer in my unit was a really nice Warrant that I got to know pretty well. He asked me to design an "A Frame" house for him and his wife, in I believe, North Carolina. I got the design almost finished when he went on a mission and got shot in the wrist and had fragments in his chin from his bullet proof vest. The flight surgeon grounded him but he went out anyway later that night, on a flare dropping mission in support of some Americans on the ground that were in danger of being overrun. Most of our ships had been shot up that day and the only one that wasn't, they selected for the mission. There was no way to install a blackout curtain behind the pilots and I told them not to use it for the mission. Better to use one of the wounded birds that could still fly. The flares are one million candlepower and the crew chief had put the seats back down to be more comfortable. One of the flares cooked off and rolled under the center seat. The crew totally blinded by that much light flew into the ground screaming and crying all the way. We were in the operations bunker and heard them go in on the radio and could do nothing but sit there and cry with them as they died. I still have the drawing of the house that he wanted. It is a reminder of the bitter times and senseless losses in war.

Not long after this I was told that my father had suffered another sever heart attack, so I came home on emergency leave once more. My father was not in a VA hospital this time as they had lost most of their funding due to political pressure from the private hospitals in the country.

After visiting with him in a Fresno hospital I talked to his doctor and he said, "Request a compassionate reassignment." Going to the Induction Facility in Fresno, I received the papers that I needed for the request and I also requested I be assigned to them for quarters and rations until new orders came down.

Chapter Eight: Fort Ord and Beyond

Being a helicopter maintenance inspector assigned to the induction center was an interesting experience. I requested a job to do so that I could keep busy and they assigned me to the medical section. There I am a helicopter mechanic assisting with pre-induction physicals? I did height, weight, blood pressure and the tests for VD. Was never comfortable with that one always wore gloves when handling their blood. Had one young man come in that had so many needle tracks on his body, I couldn't understand why he was alive. Don't think he knew either for he didn't seem to care. Then there were the ones that took drugs to try and fail their physical. Usually they would be so nervous that they would take too much and O.D. unless they could tell the doctor what they had taken there was nothing he could do to help them. The cure for one drug is fatal to another. Speaking of the doctor; a very nice young man except that he started an affair with a young woman that wanted to enlist and their first affair got her pregnant.

I got orders to Fort Ord before I found out the results of that affair. Then there was the young man that wanted to enlist and was recruited by the Marines but was grossly overweight and couldn't. He got a free physical out of the deal. We also had one young man come in with enough metal holding his leg bones together to build a small car. He also had to be rejected.

One of the highlights being there was when the Marines needed more recruits. A bunch of high school football types were sitting in a large room waiting to be told their status and in comes a Marine master Sergeant that looked like he could chew horse shoes and spit nails. He read off a list of names and then got a big grin on his face and announced; "You ladies are now drafted into the United States Marines." The bad boys all whimpered as they were led to his office to be sworn in and loaded onto buses for Camp Pendleton.

While there, the Center Commander contacted a helicopter overhaul facility in Fresno run by the National Guard. They had been repeatedly requesting a regular Army maintenance inspector be assigned to their facility, some one with combat damage experience.

Even the State Commander of the National Guard got involved and tried to get me assigned there. The Army said I was needed at Fort Ord. On arriving at Ft. Ord I found out that my MOS was not even authorized there. So much for being needed there, all of the inspectors at Ord were civilians that made at least three times as much as a Staff Sergeant in the Army.

After a few months at Fort Ord I was sent to Hunter Liggett Military Reservation South of Ord to run the Army Airfield there. Hunter Liggett was at one time one of William Randolph Hearst's ranches. On Hunter Liggett is one of the original Spanish Missions and all of the buildings that Hearst built were modeled to fit the architectural style of the mission. We had a short runway and a heliport about 200 yards from the runway. On one occasion the aviation officer from Ft. Ord flew into Liggett in a Huey and as he approached I gave him instructions to the heliport and instead he flew right down the middle of the runway.

At the same time a civilian fixed wing had received clearance to take off. The civilians do not have the radios that the Army has so I had to grab another microphone and warn him of the approaching chopper. Jumping into the airfield truck I went looking for the pilot of the chopper and missing him I radioed back to my office to get the post commander on the phone. After telling him what happened, he had the LTC standing at attention and chewed him out good for his not following landing instructions. Wish I could have been there to see that, but the airfield commander and post Executive officer, Maj Trux was and he enjoyed the show and told me about it later.

Hunter Liggett was the training facility for all Otter pilots. Otter's are a large single engine fixed wing plane built in Canada. We had several dirt airstrips for them to train on at Liggett. One time after a rain, Ord operations called and said a training flight was coming down that morning. Checking the runway that they would be using I determined that about 100 yards of the downhill portion was too wet to use. Moving the markers up to the dry portion I radioed them not to overrun to the lower part. He over ran the markers and got away with it the first time. The second landing to the lower section he applied the left brake and spun the aircraft around, burying the left wheel in the mud.

Receiving a very indignant radio call from the student pilot I got back into my truck and went out there. He expected me to tie a rope from the landing gear to my truck and pull him out while he applied full power to the Otter engine. Stupid idea and I told him so as I handed him a shovel and said dig! He was able to taxi it out after digging a long tapered trench in front of the wheel and we put a section of steel landing matt in front of it. The instructor pilot then took him back to Ord and I never saw that student again.

Sometimes if the airfield was fogged in I would call Ord and the FAA closing the field to air traffic. Then take the chain saw that I had barrowed from the engineers and go cut a pickup load of dry oak fire wood to take to my sister and brother-in-law at Bass Lake. I was allowed to cut any down dry oak and the post was full of it. Liggett had several old adobe structures left from the days of the Spanish Ranchos and other early settlers, and the ruins of an old hotel at the entrance to the post. It had been the small community of Jolon at one time, only had a small general store, gas station and trailer park left I always enjoyed exploring the old ruined Haciendas and talking to the old timers that lived there. Many of the ranchers in the area grazed their cattle on Liggett so I had to deal with their cattle on the dirt training airfields.

Hunter Liggett is now Fort Hunter Liggett with a lot of new facilities and a new longer runway about a mile from the old one. The Ft. is now an Army Reserve facility for training and has one of the best tank ranges in the Army. The dirt runways are still out in the hills but I don't know if they are used very much.

If you have seen the movie "We Were Soldiers" it was filmed in Stony Valley at Hunter Liggett and the soldier 'extra's' were California National Guard. One of the guardsmen belongs to the same American Legion post that I do in Sutter Creek, CA.

After being extended beyond the legal limit for temporary duty, orders came for me to return to Ord. Not really happy with the NCO barracks that I was assigned to and the small room that was to be my home I looked up a young man that I had served with in Vietnam and found that he and his wife lived in Castroville a few miles from Ord. They lived in a small house in a group of rentals out in the country. One of the houses was empty and needed some work. Contacting the landlord I made a deal with him to fix it up for a reduction in the rent.

While at Ord we received word that "Hanoi Jane" was going to demonstrate outside the main gate. Taking our old CH-34 helicopters and a bunch of troops we went to the high school football field and waited, hoping for a chance at her and her crowd. She was demonstrating against the military stockade as inhumane and not rehabilitating the men with whatever. The dumb broad didn't understand that all GI's have a trade and don't have to be re-trained. They are only there for punishment.

The Army decided that Ord and Fort Lewis Washington needed to have five each OH-13 instrument training helicopters for pilot training. Off we flew by commercial air to Ft. Rucker Alabama to pick up the choppers. Being the inspector on the trip I grounded all but one of them. My maintenance officer checked it over, took it for a test flight and landed in a clearing in the woods making them bring it back on a flat bed truck. I really don't know if the mechanics there could even spell the word, I do know that they were not qualified to do the job. After a week and a lot of new parts they passed inspection and we headed back west. Remember the man that I met a Ft. Eustis and visited the historical sites with? He was working in a snack bar there at Rucker. While there I was told about the local Sheriff confiscating a truckload of booze heading for the Fort. The truck and driver was returned a week or so later, the truck empty. The local law controlled all of the 'shine' in that county so the Army built a road from the airfield to main post and flew it in from then on. Fort Rucker is in a dry county. Try to imagine flying most of the way across the U.S. at about 200 feet and 60 miles an hour? Our fuel stops were sometimes on old Army airfields that had been abandoned after WW II and often we would land and before it was our turn to hover over to be refueled, we would run out of fuel. That's cutting it a bit too close. On one leg of our trip we were in a bad rainstorm and our radios stopped receiving. My pilot was a very nervous person and we started drifting south of our course to the Dallas-Fort Worth area. I located us on the map using the duel nav. Aids in the chopper but he would not believe me until we flew over a very large lake about 50 miles off our course. Almost ran out of gas getting to our next stop at a National Guard facility by Dallas. A couple times we chased coyotes as a diversion crossing the expanse of Texas, New Mexico and Arizona. The day that was to be the last in the trip for the Ft Ord group I flew in the Otter that was our chase plane and one of the choppers went down with hydraulic problems just across from the airfield in Palm Springs. The maintenance officer and I went out there, and after disconnecting all of the hydraulics we managed to hover it over a 6-foot fence and across the active runway. It took both of us using both hands to get over that fence without hydraulics. He and I stayed there for a week waiting for parts and when they came, they came with civilian mechanics that wanted to see Palm Springs.

By that Time we were both pretty well broke but we had enjoyed our stay and the good food at a little Italian place that catered to the stars. The food was so good that you went outside and walked around the block and then came back in for dessert.

Chapter Nine: Life Outside the Army and as a Reservist

Back to Ft. Ord for a few more months before getting out of the active Army I met a lady with the most beautiful legs I had ever gazed upon. I made a decision that she was the one for me and found out all I could about her. She had three children, Bill, Susan and Scott and a dog, Lucky. Took a few months to convince her but 37 years later I am glad I did. A few months after meeting Sandy I was discharged out of the active Army on the 15th of Dec 1970 and we got married on the first day of spring in 1971 in a little church at Bass Lake, CA. When we got married I wore the suit that I had made by Chan He Tailors in Singapore a few years earlier and Sandy wore a beautiful blue dress. Along with the children, both of Sandy's brothers from Portland and my step-mom, sister Connie and her husband made up the wedding party.

So ends my time in the active army and begins a period of 11 years as a civilian. In that time I worked as a carpenter. We had a son, Matthew and later we opened an antique restoration business for 12 years. One of my more interesting customers was the 'John Steinbeck House' in Salinas, repairing not only furniture but also repairs on the building. When daughter Susan got married Sandy and I made a deal with one of the local florist to duplicate a piece of furniture for him in exchange for the flowers for the wedding. After the wedding, I was concerned that they had brought more flowers than we had bargained for. Their answer was "You earned it." It is always nice to have satisfied customers. We lost our oldest son Bill in an auto accident when he was 22 years old, just before Susan graduated from Fresno State and got married. Years later, when Matthew turned 12, he and I would go camping in the Sierra's where I grew up for a week each summer. If Sandy could get the time off we would then come home and take her with us for another week. Son Scott was working for a drywall contractor in Castroville

During my time before I joined the reserves, Sandy and I became involved in a Christian retreat group that was originally started in Spain by the Roman Catholic Church to get the men back into church with their wives and children. One of the interesting men that I met at the retreat was r Col. Carl Eifler of the OSS in WWII. Carl, under the command of Major General Donovan with 12 officers and 9 NCO's went into Burma to conduct unconventional warfare. They trained the Burmese to fight the Japanese. Karl had been a police officer and border patrol agent before the war and a devout atheist. After the war he got tired of people telling him that he should become a Christian, so he went to school to prove them wrong. Studying Hebrew, Latin and Greek, he then went on to get a degree in Theology and Psychology, becoming one of the strongest Christians that I have ever met. His biography, The Deadliest Colonel, by Thomas A. Moon is a must read.

Needing medical insurance I got a job at the Monterey Vineyard Winery in Gonzales, CA as a painter and carpenter in 1982. As I was starting that job I was just completing my first year in the Army reserve assigned to the 416th Engineer Command area support center at the Presidio of San Francisco, CA.

The 416th ENCOM was tasked with maintaining reserve facilities all over the US. We had area support centers around the country in every Army Area, and at each support center there were several 5-man teams consisting of 1 Ltc, 1 Maj, 1Cpt, and 2 SFC's. My team's responsibility at times included all of Arizona, Southern Nevada, portions of the Los Angeles area, Fresno, Bakersfield, Fort Ord, Camp Roberts and Hunter Liggett. Our team, without me, even did some work for the folks in Hawaii. We at times had offices at Camp Roberts, Fort Hunter Liggett and Fort Ord. We conducted annual inspections of the physical facilities, did design and cost estimates, checked for earthquake damage and wrote procedural manuals as needed for them, got involved in environmental issues and anti-terrorism. Also we stopped some really ugly facilities from being built near the mission at Hunter Liggett by putting such stringent requirements to be compatible with the historic mission architecture.

The 416th has on a shelf at the Headquarters in Chicago, the designs for 'go to war' requirements for most of the world. They have sent teams to Haiti as well as many other areas that have seen conflict in recent years. In 1991 the regular Army Engineers failed with the engineer requirements for the Gulf War and had to apologize to the 416th and had them take over the responsibilities of airfields, roads, pipelines etc. The 416th after the Gulf War put together teams to help the Army Corps of Engineers put back together the damage done by Sadam. When asking for volunteers for that mission, Sandy told me "NO you have had your war, you don't need another one." I would have liked to have one more active duty tour before retiring.

Early in my time as a reservist it was decided that I should go to the NCO Academy at Fort Ord. Being that oldest person in the class and with the most active service time, I was made a squad leader for all the female students. Considering that most of them were not very tall I had to make allowances for them on the night cross-country land navigation coarse. We were issued a radio that did not work, so we were without communication. Having the ladies trade off carrying the 25 pound radio, we set off on a midnight adventure across the back country of Fort Ord. Having a different one of the girls in lead each 100 yards or so gave them the experience that they needed for land navigation. Moving around obstacles and such with out losing their bearings. Counting my own paces to be sure where we were. They would take from 1 ½ to 2 paces for every one of mine. We arrived late, but they found their designated point without any trouble. We only missed it by about 30 feet. The rest of the class was out looking for us in the dark, certain that we were lost. For the most part the guys in the class were a bunch of "red-neck" boys that couldn't deal with female soldiers. That must have been part of the reason that I was made the ladies squad leader. The class instructor gave me credit for the next higher NCO level academy class at graduation.

The Army had a reserve Engineer unit on the Navaho Nation that was very active helping their people with all kinds of projects on the reservation. Their reserve headquarters was in an old reconstructed portion of a historic log fort. I never got to go up there but on one trip during the gulf war a less than bright Major from our Army reserve headquarters for the Western U.S. in Los Angeles asked a Warrant Officer that was there, why his people were not doing anything for the Gulf War effort.

And he told him that they were only there as a token to appease the Indians. He was lucky not to lose his hair over that remark. The entire Engineer Reserve unit was in the gulf building roads, pipelines and airfields and what ever else was needed for the war. On the flight back from there to Fort Huachuca, AZ that less than bright person became very indignant about the way he was treated and a lady from the Huachuca engineer office tried to throw him out of the plane at 25,000 feet for his stupidity. Fortunately someone stopped her from opening the door. Unfortunately, for the Indians, their reserve unit was disbanded some time later. "Budget cuts" right. The Indians did an outstanding and professional job for the US Army and the 416th in the Gulf War.

Most of the men and women that I served with in the 416th were professional engineers, contractors and other professionals. One Commander at the Presidio of San Francisco was the Chief engineer for Cal-Trans in charge of all the bay area bridges. This was the caliber of people that most of the 416th had in its ranks also; most of those that I served with in the 416th were dedicated family men/women. Those that were not usually didn't last too long.

On one reserve facility visit, one of the reserve units was Special Forces and they had a problem with their arms storage room. They had several 'unauthorized' foreign automatic weapons stored in unsecured conditions. Our team leader redesigned the storage for them and got it modified to satisfy security needs. While there I was able to see their file on Cuba. It had dated photos of the ships coming into Cuba with the missiles on board and not showing much red paint above the water line. They also had photos of the ships leaving with the 'Missiles on board'. There was ten to twelve feet of red paint showing above the water line indicating that the missile containers were empty.

They also had recent photos of the missile sites showing that they were still maintained and ready to be used. The word was that the Russians when they left Cuba took some of the key parts and the launch codes with them, not trusting the Cubans with the missiles. I guess the only way we will ever really know is if we can get back in there officially and inspect the sites.

My last team leader had a house on a lake in Southern Monterey County that we would frequently stay at if we were not on the road, usually having BBQ Tri-Tip, Black Bean Chile, Grilled Corn on the Cob, green salad and some fine red wine. Life is tough when you have to eat our version of Army chow! At our last gathering we all brought our wives and had a farewell BBQ.

At my last parade in the 416th our Area Support Commander made 0-6 and got a Canadian band that was visiting the Navy at that location to play for us as we marched. That was something to see, a Canadian band with Bagpipes and Canadian Royal Mounted Police leading the U.S. Army onto the parade field. One of our past Commanders, a Colonel that lives in Redding took Sandy under his wing and sat in the second row with her. I was getting the Meritorious Service Medal for my time in the 416th and he felt that she should have a seat close to the front. The front row was reserved for the family of the new Colonel.

After I spent 18 years at the winery expanding into metal fabricating, farming, plumbing, design and construction among other things Sandy and I decided to retire and move from the home she owned when we met. This also concluded 18 years in the Army Reserve as a construction inspector/supervisor.

Chapter Ten: Finally, Retirement

In 1999 at age 60, and 27 years total Army service, I was retired mandatorily form the U.S. Army on my birthday. In April 2000, I retired from the winery and in May we found a new home in the mountains above Jackson, CA. Here we are active in helping to start a new Lutheran church in Jackson and we are active in the Mother Lode Rose Society and the Sierra Madre Garden Club. Sandy is able for the first time to garden and grow roses and we can now enjoy our grown children, grandchildren and oh yes, by now, in 2007, we have one great grandchild a beautiful little girl.

Is my story unusual? I don't think so, I have just been fortunate to be in the right places at the right time for most of my life so that I could help others and I hope to be good influence on some of the people that I have met. Some say that I was lucky. I don't think luck had anything to do with it. Not being in the infantry certainly helped but lucky, no. Luck is something that you need in gambling not life. I learned a long time ago that gambling is not good and certainly not acceptable to the Lord. The Lord knows that I am still struggling with my faith. Traveling in the Army taught me that the world is full of new friends that I have not met yet and foods that need to be sampled. Why have I survived? Only God knows and He has not openly shared that with me. Early in this story I told you about Christian upbringing and my guardian angel. I feel that it was in God's plan that Sandy and I meet, marry have a child and to raise our children and for us become active in His church in Salinas; Our Savior Lutheran. We then retired and moved, and are helping with His church here in Jackson, CA. He has brought me through some near fatal accidents and incidents in and out of the Army over the years and an 80% blockage to one of the arteries on my heart. I thank God for his continuing forgiveness and praise Him from whom all blessings flow.

I have not recalled a lot of "war" stories, as are depicted in most of the movies. I believe that it is important to recall the people that I have met in my travels in the Army and the places as real places and not just battles won or lost. As a soldier I have always tried to do what was expected of me and a little more. My father's advice from so long ago made a good impression on me and on my actions as a United States Soldier. So many of the movies made about the Vietnam war are Just "Blood and Guts" fiction and never even try to show the people over there as anything but poor and backwards buffoons in rice paddies, ignoring the ancient cultures and needs of the Vietnamese people. American GI's are so often depicted as drug crazed alcoholics and not the young confused soldiers that most of them were, trying to do the job that they were trained to do, often without really understanding why.

In closing, for my family this is a brief history of my time and the awards given me.

Total Awards and Decorations:
Distinguished Flying Cross
Meritorious Service Medal
Air Medal (16)
Army Commendation Medal
National Defense Service Medal
Armed Forces Expeditionary Medal (Domi-nica)
Good Conduct Medal (2)
Army Reserve Components Achievement Medal (3)
Armed Forces Reserve Medal
Vietnam Campaign Medal (7)
NCO Development (2)
Army Service Ribbon
Republic of Vietnam Service Medal
Presidential Unit Citation
Meritorious Unit Citation
Vietnam Cross of Gallantry w/Palm Devise
Aircraft Crewman's Badge
Air Assault Badge (original 11th Air Aslt Div)
Expert Rifle/Revolver

I sit here, now thinking about being a soldier then and our current armed forces. The youngsters now are so lucky that most of the country backs them and the job they are doing. Their equipment is superior to what we had, but it still takes soldiers on the ground to finish the job. That will never change. We had compasses, maps and U. S. Mail. Now they have GPSs, e-mail and cell phones, but they do not have Bob Hope and his merry traveling companions.

Being active in Post 108, American Legion Honor Guard brings great satisfaction of being able to give back to our departed veterans the honor due them for their sacrifices. It is so little that we are able to give. To be able to ceremonially fold an American Flag, fire a gun salute, play taps and present to the next of kin the folded flag and a few mementoes of the service provided.

Made in the USA
San Bernardino, CA
14 September 2014